T0319675

Sensor Data Analysis and Management

Sensor Data Analysis and Management

The Role of Deep Learning

Edited by

A. Suresh
Department of Networking and Communications, SRM Institute of Science and Technology, Tamil Nadu, India

R. Udendhran
Department of Computer Science and Engineering, Sri Sairam Institute of Technology, Tamil Nadu, India

M.S. Irfan Ahmed
Department of Computer Science and Information, Taibah University, Al-Ula Campus, Madhina, Saudi Arabia

This edition first published 2021
© 2021 John Wiley & Sons, Ltd.

All rights reserved. No part of this publication may be reproduced, stored in a retrieval system, or transmitted, in any form or by any means, electronic, mechanical, photocopying, recording, or otherwise, except as permitted by law. Advice on how to obtain permission to reuse material from this title is available at http://www.wiley.com/go/permissions.

The right of A. Suresh, R. Udendhran, and M.S. Irfan Ahmed to be identified as the authors of the editorial material in this work has been asserted in accordance with law.

Registered Offices
John Wiley & Sons, Inc., 111 River Street, Hoboken, NJ 07030, USA
John Wiley & Sons Ltd, The Atrium, Southern Gate, Chichester, West Sussex, PO19 8SQ, UK

Editorial Office
The Atrium, Southern Gate, Chichester, West Sussex, PO19 8SQ, UK

For details of our global editorial offices, customer services, and more information about Wiley products, visit us at www.wiley.com.

Wiley also publishes its books in a variety of electronic formats and by print-on-demand. Some content that appears in standard print versions of this book may not be available in other formats.

Limit of Liability/Disclaimer of Warranty
While the publisher and authors have used their best efforts in preparing this work, they make no representations or warranties with respect to the accuracy or completeness of the contents of this work and specifically disclaim all warranties, including without limitation any implied warranties of merchantability or fitness for a particular purpose. No warranty may be created or extended by sales representatives, written sales materials, or promotional statements for this work. The fact that an organization, website, or product is referred to in this work as a citation and/or potential source of further information does not mean that the publisher and authors endorse the information or services the organization, website, or product may provide or recommendations it may make. This work is sold with the understanding that the publisher is not engaged in rendering professional services. The advice and strategies contained herein may not be suitable for your situation. You should consult with a specialist where appropriate. Further, readers should be aware that websites listed in this work may have changed or disappeared between when this work was written and when it is read. Neither the publisher nor authors shall be liable for any loss of profit or any other commercial damages, including but not limited to special, incidental, consequential, or other damages.

Library of Congress Cataloging-in-Publication Data
A catalogue record for this book is available from the Library of Congress

Hardback ISBN: 9781119682424; ePub ISBN: 9781119682486; ePDF ISBN: 9781119682455;
oBook ISBN: 9781119682806.

Cover image: © Issaro Prakalung/EyeEm/Getty Images
Cover design by Wiley

Set in 9.5/12.5 and STIXTwoText by Integra Software Services, Pvt. Ltd, Pondicherry, India
Printed and bound by CPI Group (UK) Ltd, Croydon, CR0 4YY

C9781119682424_011121

Contents

About the Editors

Dr A. Suresh, B.E., M.Tech., PhD, works as Associate Professor, Department of Networking and Communications, School of Computing, Faculty of Engineering and Technology, SRM Institute of Science and Technology, Tamil Nadu, India. He has nearly two decades of experience in teaching and his areas of specializations are Data Mining, Artificial Intelligence, Image Processing, Multimedia and System Software. He has published two patents and 90 papers in international journals. He is author of the book *Industrial IoT Application Architectures and Use Cases* published by CRC Press and editor of *Deep Neural Networks for Multimodal Imaging and Biomedical Application* published by IGI Global. He has published more than 40 papers in national and international conference proceedings. He is a member of IEEE (Senior Member), ISTE, MCSI, IACSIT, IAENG, MCSTA and a Global Member of Internet Society (ISOC). He has organized several national workshops, conferences and technical events. He is regularly invited to deliver lectures at various programs for imparting skills in research methodology to students and research scholars. He has published four books with Indian publishers in the fields of Hospital Management, Data Structures & Algorithms, Computer Programming, Problem Solving and Python Programming and Programming in "C". He has hosted two special sessions for IEEE-sponsored conferences in Osaka, Japan and Thailand.

Mr R. Udendhran, B.Tech, M.Tech. (PhD), works as Assistant Professor Grade III, Department of Computer Science and Engineering at Sri Sairam Institute of Technology, Tamil Nadu, India. He is a computer science research scholar focusing on Deep Learning. He worked as a data scientist and has presented research work at an international conference held at the University of Cambridge (available in the ACM Digital Library) and has published approx. 5 research papers indexed in the Web of Science and 11 research papers in the Scopus database. He is a member of ISTE, MCSI, IACSIT, and IAENG.

Dr M. S. Irfan Ahmed is working as Associate Professor in the Department of Computer Science and Information, Faculty of Science and Literature at Taibah University, Saudi Arabia. He is a member of ISTE, MCSI, IACSIT, and IAENG.

List of Contributors

Aditya Patel
LNCT College, Bhopal, India

Ajmi Nader
Micro-Optoelectronic and Nanostructures Laboratory, University of Monastir, Faculty of Sciences of Monastir, Tunisia

Akash Saxena
CITM, Jaipur, India

Akhilesh Vikas Kakade
SAP Labs, Bangalore, India

G.R. Anantha Raman
Professor, MRIET,
Secunderabad, Telangana, India

M. Balasaraswathi
Associate Professor, ECE,
Saveetha School of Engineering,
SIMATS, Chennai, India

G. Deivendran
National Engineering College,
Tamil Nadu, India

Ganesan Sivarajan
Associate Professor, Department of Electrical and Electronics Engineering, Government College of Engineering Salem, Tamil Nadu, India

Hariprasath Manoharan
Assistant Professor, Department of Electronics and Communication Engineering, Audisankara College of Engineering and Technology, Gudur, Andhra Pradesh, India

Helali Abdelhamid
Micro-Optoelectronic and Nanostructures Laboratory, University of Monastir, Faculty of Sciences of Monastir, Tunisia

S. Joseph Gladwin
Associate Professor, ECE,
SSN College of Engineering, Chennai, India

K.M. Karthick Raghunath
Associate Professor, MRIET,
Secunderabad, Telangana, India

Mghaieth Ridha
Micro-Optoelectronic and Nanostructures Laboratory, University of Monastir, Faculty of Sciences of Monastir, Tunisia

T.J. Nagalakshmi
Assistant Professor, ECE,
Saveetha School of Engineering,
SIMATS, Chennai, India

B. Paramasivan
National Engineering College,
Tamil Nadu, India

S. Pravin Kumar
AI Engineer, Smartail Pvt Ltd,
Chennai, India

Prisilla Jayanthi
The Airports Authority of India Ltd, India

L. Priya
Department of IT, Rajalakshmi
Engineering College,
Chennai, India

Radhika Baskar
Associate Professor, ECE,
Saveetha School of Engineering
SIMATS, Chennai, India

S. Rajkumar
Vellore Institute of Technology,
Vellore, India

L. Ramanathan
Vellore Institute of Technology,
Vellore, India

D. Ravikumar
Professor, ECE,
Vel's University, Chennai, India

V. Saravanan
Department of Computer Applications (PG), Dr. SNS Rajalakshmi College of Arts and Science, Coimbatore, India

Ms. A. Sathya
Department of IT, Rajalakshmi
Engineering College,
Chennai, India

N. Shanmuga Priya
Department of Computer Applications (PG), Dr. SNS Rajalakshmi College of Arts and Science, Coimbatore, India

V. Sivasankaran
Sreenivasa Institute of Technology and Management Studies, Chittoor, Andhra Pradesh, India

Subramanian Srikrishna
Professor, Department of Electrical and Electronics Engineering, Annamalai University, Chidambaram, Tamil Nadu, India

Sudeep Ray Gaur
LNCT College, Bhopal, India

K. Suganthi
Vellore Institute of Technology,
Vellore, India

A.S. Syed Fiaz
Assistant Professor, CSE,
Vel Tech Rangarajan Dr. Sagunthala R&D
Institute of Science & Technology,
Chennai, India

S. Thanga Revathi
Department of IT, Rajalakshmi Engineering College, Chennai, India

G. Uganya
Assistant Professor, ECE,
Saveetha School of Engineering,
SIMATS, Chennai, India

N. Vijayaraj
Assistant Professor, CSE,
Vel Tech Rangarajan
Dr. Sagunthala R&D Institute of Science & Technology, Chennai, India

Vikram Rajpoot
GLA University, Mathura, India

S. Vimal
National Engineering College,
Tamil Nadu, India

S. Vishal Balaji
National Engineering College,
Tamil Nadu, India

Preface

Sensor Data Analysis and Management: The Role of Deep Learning delivers an insightful and practical overview of the applications of deep learning techniques to the analysis of sensor data. The book collects cutting-edge resources into a single collection designed to enlighten the reader on topics as varied as recent techniques for fault detection and classification in sensor data, the application of deep learning to Internet of Things sensors, and a case study on high-performance computer gathering and processing of sensor data.

The editors have curated a distinguished group of perceptive and concise papers that show the potential of deep learning as a powerful tool for solving complex modelling problems across a broad range of industries, including predictive maintenance, health monitoring, financial portfolio forecasting, and driver assistance.

The book contains real-time examples of analyzing sensor data using deep learning algorithms and a step-by-step approach for installing and training deep learning using the Python Keras library. Readers will also benefit from the inclusion of:

- A thorough introduction to the Internet of Things for human activity recognition, based on wearable sensor data
- An exploration of the benefits of neural networks in real-time environmental sensor data analysis
- Practical discussions of supervised learning data representation, neural networks for predicting physical activity based on smartphone sensor data, and deep-learning analysis of location sensor data for human activity recognition
- An analysis of boosting with XGBoost for sensor data analysis

Perfect for industry practitioners and academics involved in deep learning and the analysis of sensor data, *Sensor Data Analysis and Management: The Role of Deep Learning* will also earn a place in the libraries of undergraduate and graduate students in data science and computer science programs.

Chapter 1 presents an effective better resource allocation method by using a multilayer neural network named RBNNOM. Additionally, the novel method of RBNNOM considers priority-based resource allocation in terms of qualities and quantities.

Chapter 2 demonstrates recognition strategy that can identify body sensor data irrespective of the cell phone's position. An area for potential research would be to find novel methods of movement recognition. Future research could also concentrate on recognizing new exercise routines trying to gather information from more users of different ages; and removing highlights that could fine-tune the segregation of various exercises.

Chapter 3 emphasizes the need for secure cyber security systems for sensor data. In recent years, the growth of network-based services has been incredible. Therefore, reliable network security becomes of primary importance in the cyber world.

Chapter 4 presents the challenges and risks of routing sensed data or newly generated events in a critical environment, which is always of major interest for researchers. A wireless sensor network deployed in critical infrastructure/environs mostly implies the difficulties of processing the sensed data, which may raise problems for compatibility factors in network deployment – for example, jeopardizing scenarios and monitoring emergencies such as fire eruption and explosion and environment-oriented and hazardous pollution.

Chapter 5 focuses on the potential of student motion behavior analysis. This study is conducted for the learning of repeated motion behavior with respect to the students and thereafter to show that it is possible to detect unusual behavior using the knowledge of frequent behavior. The best example for this scenario is taking a wrong route and getting lost. An important objective of pervasive computing is to give accurate information about human behavior. It has a wide range of applications such as in medicine, security solutions, and student monitoring in educational campuses.

Chapter 6 considers improving the working efficiency of sensors and for testing them under different conditions, where a predictive algorithm is essential. This is possible only when deep learning methods are used, whereby different strategies are followed when problems occur on the network. If sensors are installed, then the network depends on main node for the purpose of storing and accessing the data. For sensing the information and sending it to applications, such as those for health monitoring, there should be less delay. This robust prediction of health is necessary because it can save lives.

Chapter 7 proposes a hybrid algorithm based on KNN and QPSO for improving WSN lifetime by updating the location of the BS to reduce long-distance communication between the BS and sensor nodes. QPSO is applied to optimize KNN. Furthermore, the fitness function is also designed considering two parameters – energy consumption and distances

Chapter 8 discusses the EHR and the various sensors used in healthcare systems. The use of sensor data and its parameters is elaborated. The various feature extraction techniques are discussed, and case studies are provided for better understanding feature extraction through sensor data.

Chapter 9 proposes an approach that successfully handles object detection problems, which significantly improves objection detection in satellite images using modified pyramid scene parsing networks. This work incorporates several steps such as the adaptation of fully convolutional networks to multispectral satellite images and the evaluation of several data fusion strategies on semantic segmentation tasks of satellite images with a combined training objective.

Chapter 10 focuses on improving heart disease prediction accuracy by using different machine learning algorithms. Neural networks have proven to be more efficient in prediction and can be used for classification. Feature selection techniques can be created to obtain a more extensive view of the critical highlights to build the presentation of coronary illness forecast.

Dr. A. Suresh
Mr. R. Udendhran
Dr M. S. Irfan Ahmed

1

Efficient Resource Allocation Using Multilayer Neural Network in Cloud Environment

N. Vijayaraj[1], G. Uganya[2], M. Balasaraswathi[3], V. Sivasankaran[4], Radhika Baskar[3], and A.S. Syed Fiaz[1]

[1] *Assistant Professor, CSE, Vel Tech Rangarajan Dr. Sagunthala R & D Institute of Science & Technology, Chennai*
[2] *Assistant Professor, ECE, Saveetha School of Engineering, SIMATS, Chennai*
[3] *Associate Professor, ECE, Saveetha School of Engineering, SIMATS, Chennai*
[4] *Sreenivasa Institute of Technology and Management Studies, Chittoor, Andhra Pradesh*

1.1 Introduction

Nowadays, due to the development of the cloud computing environment, most information and communications technology (ICT) players have moved to new product management and application models—for example, Apple iCloud, Google App Engine, Amazon EC2, IBM Cloud, VMware Cloud, etc. Cloud computing is an important emerging field in ICT, making people's life easier by increasing productivity and processing speed, reducing cost and time consumption, facilitating backup and storing of multiple data, and enabling automation in the distribution of products and in development. It is quite challenging to offer trustworthy, powerful, qualitative, and cost-effective cloud services. In distributed cloud environments, a large number of dynamic resources are circulated around the world. Hence, the allocation of resources between the cloud user and cloud provider is complex, and cloud providers should be able to manage their resources with QoS and maximum customer satisfaction. Inadequate resource allocation leads to poor quality, bad performance, and substandard customer satisfaction, all falling below the criteria specified in service-level agreements (SLAs). Therefore, efficient and heterogeneous resource allocation is essential to avoid these problems.

In previous studies, the resource allocation problem was solved based on two methods: (i) reactive method, and (ii) proactive method. The reactive method is a common method in ICT, but it is not considered an effective method. The proactive method was developed to improve the performance of the system by allocating resources in a predefined manner. However, proactive-based methods, including time series (TS), queuing theory (QT), and reinforcement learning (RL), have some limitations. These include numerous data in TS,

Sensor Data Analysis and Management: The Role of Deep Learning, First Edition. Edited by A. Suresh, R. Udendhran, and M.S. Irfan Ahmed.
© 2021 John Wiley & Sons, Ltd. Published 2021 by John Wiley & Sons, Ltd.

reconstruction of architecture when changing resources in QT, and large time requirements in RL. To resolve these proactive method constraints, feedback-based approaches have been introduced in difficult computing systems.

These feedback-based mechanisms are of two types, based on the allocation of resources to cloud services. These are (i) single input and single output (SISO), and (ii) multiple input and multiple output (MIMO). SISO is developed only for providing a single kind of resource allocation. However, in distributed cloud environments, users and providers need a mixture of resource allocation, and this leads to the violation of SLAs. To overcome this SISO limitation, the MIMO feedback control system was developed for multiple groupings of resource allocation by combining multiple numbers of separate SISO feedback control systems. But this type of MIMO feedback control system also leads to poor QoS and SLA violations. To overcome this issue, a coordinated multiple input multiple output feedback system was developed, which enhances the QoS by combining all the inputs in an integrated manner. Depending on the amount of work given to the services, the resources are allocated in a synchronized manner.

Other existing work focuses on adaptive multivariable resource mechanisms for multiple-type resource allocation with respect to cloud users and cloud providers. However, this has several limitations including scalability, reduced performance of allocated resources, and critical QoS.

To overcome these problems, we propose a better resource allocation system by using the RBNNOM multilayer neural network. Additionally, this novel method of RBNNOM considers priority-based resource allocation in terms of qualities and quantities and is responsible for QoS with good learning capabilities. Finally, this proposed work analyzes node prices and priority weights for cloud users and cloud providers in the cloud environment.

1.2 Related Work

Cloud computing is used in many applications including industrial, healthcare, environmental, and public science domains. However, nowadays, resource allocation is the key research area. Resource allocation is limited by factors such as bandwidth and bulk traffic. Yu et al. discussed a link-mapping algorithm for splitting the multiple paths and a migration algorithm [1] for relocation of paths in the virtual network (VN). This work is mainly proposed for reducing bandwidth requirements. Wei et al. proposed two methods for resource allocation based on game theory. These methods are binary integer (BI) programming method [2] and evolutionary mechanism (EM). In the first method, each and every node resolves the resource optimization problem separately, but this method is not suitable for multiple-resource tasks. Therefore, the second method has been designed additionally considering the multiplexing of resource assignments and is also used to reduce productivity loss. Yang et al. proposed a profile-based method for tackling the scalability problem in resource allocation [3]. The selection of profile is based on the application of the cloud environment. The resource is distributed in three ways. These are predistribution, distribution, and postdistribution.

Chen et al. designed a virtualization model and heuristic resource combination algorithm (HRCA) that is used to transfer physical type of resources to logical resources [4]. In this work, HRCA consists of two algorithms. These are matching and reconfiguration algorithms for allocating dynamic resources. Ramachandran et al. proposed two types of tenant models: the tenant requirement model (TRM) for measuring the functional requirements of tenants, and the tenant provider model (TPM) for allocating resources depending on dissimilar tenant information [5]. Wang et al. proposed a combined knowledge representation based on basic design, process, and product information [6]. In this work, the information structure is developed by four methods: filtering the information, summarizing the information of nodes, determining the group of nodes, and finding the solutions for difficult questions related to dynamic resource allocation.

Farahnakian et al. developed the ant colony system–virtual machine consolidation (ACS-VMS) for dynamic resource allocation [7]. This work is mainly used for reducing energy consumption when distributing resources to cloud users, as well as for reducing violations of SLA. Violations of SLA are mainly caused by migration and the overuse of resources. Sim proposed the focused selection contract net protocol (FSCNP) for the automatic collection of cloud services and service capability table (SCT) for recording the cloud services [8]. In this work, the author considers only the cost consumption, but it is not suitable for QoS and the cooperation of time slots.

Papagianni et al. discussed a unified resource allocation framework [9] that has two phases for the problem of mixed integer programming (MIP). These are the node mapping phase (NMP) and the link mapping phase (LMP), which provide a solution for the resource flow distribution problem. Linlin Wu et al. proposed a customer-driven SLA [10] that is mainly used for improving user satisfaction. This work is mainly used to reduce costs and SLA violations in software as a service (SaaS). However, this work needs further improvement in multitier applications. Tao et al. discussed the case library and Pareto-solution-based genetic algorithm (CLPS-GA) for heterogeneous resource allocation [11]. In this work, the authors mainly considered two types of programming information, including processor information and user requests. Peng et al. proposed a radial basis function neural network (RBFNN), based on multi-objective genetic algorithm [12], for optimum resource allocation. But, in this work, the authors did not consider the scalability, node price, and priority load.

In cloud environments, the allocation of resources is broadly classified into two methods: (i) reactive method, and (ii) proactive method. In reactive method, the resources are allocated after getting the threshold values. However, proactive method is preferred to reactive method because, in proactive method, resources are allocated in a predefined manner, which is used to mitigate traffic in complex systems.

1.2.1 Reactive Methods for Resource Allocation

Numerous methods are proposed to overcome the issues in resource allocation based on the reactive approach. Xiao et al. introduced virtual machine (VM) live migration technology for resource allocation depending on the applications. Also, the authors proposed the skewness method [13, 14], which is used to compute the equality of the resource utilization. But, in this method, combinations of resource allocations are not possible. Wang et al. proposed an energy-efficient VM assignment [14], which can be used to decrease traffic and active switches in data center networks (DCNs). In these two methods, single-type resource allocation is not possible.

Kumbhare et al. introduced the GA-based VM agreement, which was proposed to find the solutions for optimization problems [15]. In this work, the authors proposed two types of experiments, that is, centralized and shared. These methods depend on bin packing algorithm and give better solution for the optimization problem. However, it is suitable only for single-type resource allocation, and it leads to SLA violations. Mashayekhy et al. proposed an auction-based online mechanism [16] for resource allocation and pricing. In this study, the authors discussed three processes—creating, distributing, and pricing the numerous resources. Nevertheless, this mechanism too cannot be used for different types of resource allocation and combinational resource allocation.

Zhang et al. proposed the compatible online cloud auction (COCA) mechanism [17]. It is possible to use this mechanism for combinational resource allocation, although this is not suitable for multiple-type resource allocation. Zhang et al. proposed the C3 mechanism [18], which includes three features. These features are prediction, association, and relocation. The autoregressive integrated moving average (ARIMA) method is used for prediction, and the live relocation scheme is used for migration. However, this work is not developed for combinational resource allocation, and it leads to SLA violations.

From the preceding analysis, several reactive methods were developed for dynamic allocation of resources. However, there are still some limitations. Achieving high- and low-level limits is challenging, and it is difficult to respond within the appropriate time frame, since these approaches are unpredictable (see Table 1.1).

1.2.2 Proactive Methods for Resource Allocation

Many methodologies are developed to overcome the issues in proactive-based resource allocation. Calheiros et al. proposed the proactive ARIMA model based on prediction [19]. This work is particularly developed for SaaS applications by gathering the resource analyzer element. However, this work is not suitable for multiple resources and combinational resources, and it leads to SLA violations. Morshedlou and Meybodi proposed learning automation (LA)-based proactive resource allocation [20], which was mainly introduced for reducing SLA violations. In this work, the authors introduced a new feature called *willingness to pay* (WTP) for the facilities, mainly for customer happiness by providing satisfaction. Nevertheless, this work also has some deviations, including SLA violations for multiple types of resources.

Yan et al. proposed trust management based on access control (AC) proactive method [21]. In this study, resource access is controlled by a trust-based proactive system. Also, the cloud service is protected by an automatically generated encrypted key. Similarly, Ali et al. proposed secure data sharing in cloud (SeDaSC) based on access control proactive method [22]. This work is mainly introduced for cloud computing in mobile applications. However, these two access control proactive mechanisms are not successful due to SLA violations. Tolosana-Calasanz et al. discussed an autonomic controller based on a QT proactive approach [23] in a comet cloud environment. In this study, resource allocation is based on resource types, catastrophe rate, and implementation cost. This work is only developed for single-type resource allocation, and this work is also not suitable for combinational resource allocation. Gai et al. introduced the dynamic data allocation advance (D2A)

Table 1.1 Summary of existing solutions based on reactive method.

S. no.	Year	Reference no.	Methodology	Types of resource allocation	Combinational resource allocation	Avoiding of SLA violation	Issues
1.	2013	[1]	Skewness algorithm with VM live migration	M	No	No	1) Difficult to get higher- and lower-level limits 2) Responding within time is difficult 3) Unpredictable approaches
2.	2013	[2]	Bidding language and COCA mechanism	S	Yes	No	
3.	2014	[3]	Energy-efficient VM assignment	S	Yes	No	
4.	2015	[4]	Genetic algorithm (GA)-based VM assignment	S	Yes	No	
5.	2015	[5]	Auction-based online mechanism	S	No	Yes	
6.	2016	[6]	COCA mechanism	S	Yes	No	
7.	2018	[7]	Prediction, consolidation, and migration (C3) model	M	No	No	

S = single-type resource allocation; M = multiple-type resource allocation prediction.

algorithm based on an RL proactive mechanism [24] for multiple-type resource allocation. However, this work is not preferred for combinational resource allocation.

From the preceding analysis, it is clear that the existing proactive-based methods require more time to respond, and it is hard to redesign them when changing the capabilities of the system. The researchers then tried to implement a control-theory-based proactive method for providing automatic resource allocation. The feedback control structure is mainly classified into SISO and MIMO (Table 1.2).

1.2.3 Control-Theory-Based Proactive Method

Rao et al. proposed a self-tuning fuzzy-control scheme for multiple types of resources [25]. Also, the authors proposed a two-layer dynamic QoS framework; one layer is for the

Table 1.2 Summary of existing solutions based on proactive method.

S. no.	Year	Reference no.	Methodology	Types of resource allocation	Combinational resource allocation	Avoiding of SLA violation		Issues
1.	2014	[8]	ARIMA model based on prediction	S	No	No	Time series (TS)	Mostly based on past collected data
2.	2014	[9]	WTP model based on LA	S	Yes	No	Learning auto-mation (LA)	
3.	2015	[10]	Trust management framework	S	Yes	No	Access control (AC)	More time required to react
4.	2015	[11]	SeDaSC (Secure data sharing in cloud)	S	Yes	No		
5.	2016	[12]	Autonomic controller based on QT	S	No	No	Queuing theory (QT)	Recom-putation is difficult
6.	2016	[13]	D2A (dynamic data allocation advance) algorithm	M	No	Yes	Reinfor-cement learning (RL)	More time taken for learning

S = single-type resource allocation; M = multiple-type resource allocation prediction.

distribution of resources, and the other one for the differentiation of facilities. But this work is not perfectly used for combinational resource allocation and leads to the violation of SLAs. Islam et al. introduced the water-constrained workload scheduling (WATCH) algorithm for same types of resource allocation [26], which dynamically allots the loads to the data center (DC). This work too is not suitable for combinational resources. Gong et al. discussed proportional integral derivative (PID) controllers self-tuning mechanism, based on neural network radial basis function (RBF), for distributing the resources [27]. In this work, the authors used this method for three types of assignments in the health industry—health center, patient's room, and emergency. The authors mainly focused on emergency assignment. However, this method too is not suitable for combinational resource allocation [28] and SLAs. Song et al. discussed the self-adaptation framework, which has two processes for service-oriented manufacturing cyber-physical system (SMCPS) [28]. One is the workflow (WF) nets, and the other is the level of optimization model for quality of local and universal data. This method is not implemented for dynamic control strategies and combinational resources. Baresi et al. proposed the autoscaling mechanism gray-box distinct-time feedback controller based on the management of resources [29]. From the preceding analysis, the SISO method is mainly used for single-type resource allocation and is not suitable for combinational resource allocation. The SISO method then leads to dissimilarity problems in the allocation of resources.

The dissimilarity problem is avoided by allocating multiple types of resources—such as CPU and memory—using the MIMO feedback control structure. Xia et al. proposed an analytic-model-based MIMO approach [30] for similar resources. It leads to the state eruption problem, that is, the number of states exponentially increases with the size of the problem because of scalability issues. Farokhi et al. proposed a fuzzy-control-based autonomic resource controller [31] for multiple combinational resources. In this work, the authors discussed three controllers, including CPU, memory, and fuzzy controller (FC). The FC controls the other two controllers. The CPU and memory controller distribute the resources depending on the input given by the FC. However, this method violates the SLA. Saikrishna et al. proposed a linear parameter varying (LPV)-framework-based MIMO control structure [32]. In this work, the authors managed the VM resources by neglecting new VM resources during the process of passing the assignments. This method allocated multiple types of resources, but not combinational resources. Gong et al. discussed the adaptive resource allocation [33] for service-based systems by distributing single resources to numerous numbers of servers. In this work, the authors used the three kinds of elements—the first element is *smoothing*, which is used to calculate the assignments to each server; the second element is *optimization*, which is used to calculate the distributing resources; and the third element is the *system*, which is used to fine-tune the controller parameter depending on past and present data. So, this method is not suitable for multiple types of resources and combinational resources. Gong et al. proposed the adaptive MIMO feedback control mechanism [34], which provides multiple types of resources to numerous numbers of servers. In this work, the authors introduced the method for combinational resource allocation and utilization of resources. However, this work is not suitable for large number of cloud users and providers, as presented in Table 1.3.

1.3 Data-Based Task Construing

The process of resource allocation in a distributed environment and its interactive system enable the growth of the multitenant system. The multitenant system works for various cloud users with dissimilar requirements within a cloud environment. This system contains different cloud users who segment the single-instance database and application in the cloud environment to finish the same job. In the SLA, the system offers a framework among the cloud user, cloud service provider, and multitenancy [35]. The common process of resource allocation based on multilayer neural network and optimization model is shown in Figure 1.1.

Table 1.3 Summary of existing solutions for control-theory-based proactive method.

S. no.	Year	Ref. no.	Methodologies	Types of resource allocation	Combinational resource allocation	Avoiding of SLA violation	Feedback control structure
1.	2013	[14]	Self-tuning fuzzy control (STFC)	M	No	No	SISO
2.	2015	[14]	Water-constrained workload scheduling (WATER)	S	No	No	
3.	2015	[15]	PID self-tuning controller	S	No	No	
4.	2016	[16]	Self-adaptation framework for SMCPS	S	No	No	
5.	2017	[17]	Autoscaling mechanism	S	No	No	
6.	2013	[18]	Stochastic modeling	M	No	No	MIMO
7.	2015	[19]	Autonomic-resource-controller-based fuzzy control	M	Yes	No	
8.	2016	[20]	Linear parameter varying (LPV) framework	M	No	No	
9.	2017	[21]	Feedback control for service-based systems (SBSs)	S	No	No	
10.	2019	[22]	Adaptive multivariable control scheme	M	Yes	Yes	

S = single-type resource allocation; M = multiple-type resource allocation prediction.

1.3.1 Data Model

In the cloud computing environment, the data model process is complex, as the product is disintegrated into self-determining the subtasks that have the need of different developers for collaboration. Commonly, the different types and qualities of the resources distributed to subtasks which are determined by the developers and referred by the data model are based on the domain. Still, this type of data model is difficult to classify for efficient and effective reuse of the data. In the dynamic request of services, an RBNNOM model is

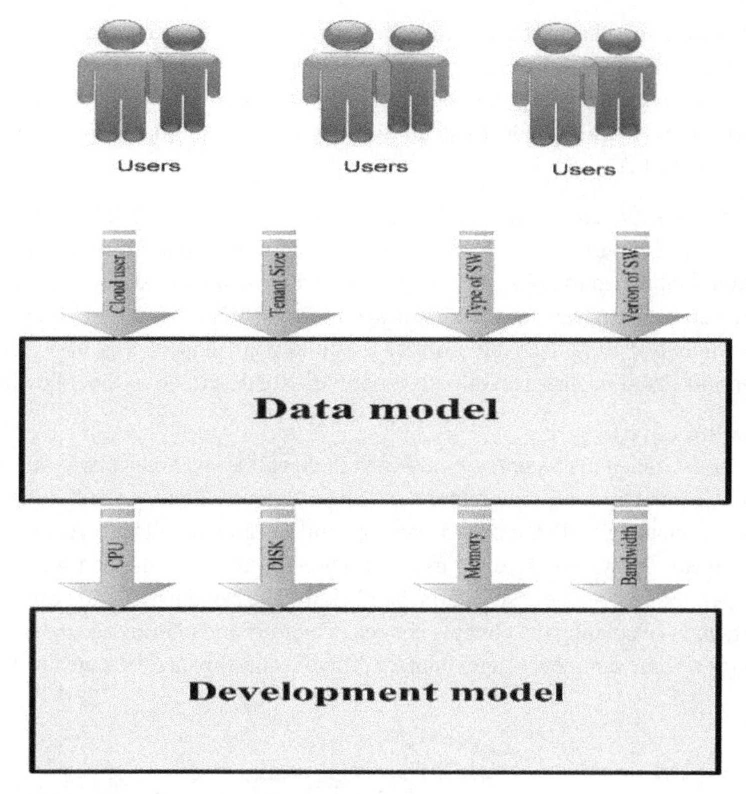

Figure 1.1 Architecture of resource allocation.

developed for understanding the job based on the data model, which transforms the requirement of the tenant and the details of the quality of the resource.

1.3.2 Development Model

In a cloud computing environment, resource allocation is highly heterogeneous with an effective development model that is recognized as standard in view of tenant priority, energy consumption, and load balancing. In the direction of this framework using the cloud simulation environment, all the algorithms and system models are encapsulated and established through a middleware solution and growth of the whole software system.

1.3.3 A Data Model Established by the RBNNOM

In a cloud computing simulation platform, the application is described in detail, and the three types of jobs are addressed and named as memory-constrained jobs, CPU-constrained jobs, and I/O-constrained jobs.

1) Memory-constrained jobs frequently involve several business databases, system models, and files, and are assigned to the nodes with huge memory sizes.

2) CPU-constrained jobs are best shared for the processing of parallel computing jobs or general scientific calculations based on the domain.
3) I/O-constrained jobs incorporate best quantity of data intensive for exchanges. In this situation, the general speed of calculation is partial by the data bandwidth between memory and CPU.

The interactions between job types and resource attributes are significantly nonlinear to its common math model unable to satisfy the requirement of the cloud user. In this chapter, RBNNOMis established to define the relationships between address and various common factors. The structure of the network is shown in Figure 1.2; it has a four-layered feed-forward layer network and effectively notifies the nonlinear models. The basic functions of each layer and signal mechanism of neural network are described in the following text.

Contribution Layer

This layer is also called the *input layer*, and each node of this layer matches to one input variable. It identifies the resource requirement of each job, and this input vector is initiated by four variables—the cloud user ID, tenant size, version, and category of software. A cloud user commonly completes the types of jobs, for example, based on the domain and user requirement, the developer, the charge, the cost, and the bandwidth. The cloud user ID and category of software can help in calculating the possible behaviors of users and offering accurate resources to the cloud user. Some samples of input data for different software are presented in Table 1.4.

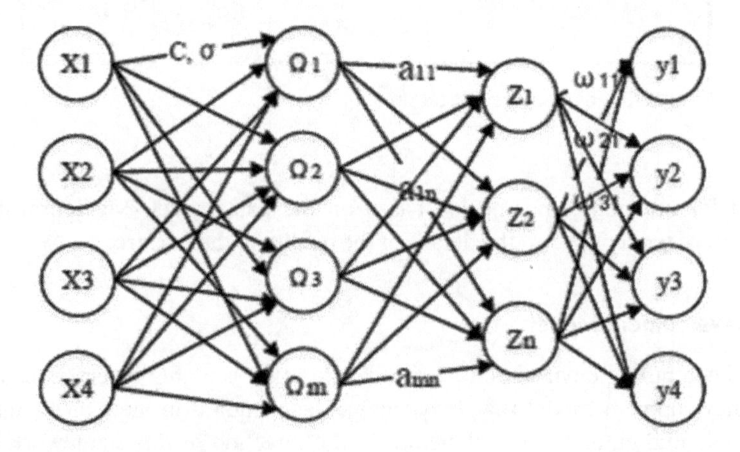

Figure 1.2 Structure of the RBNNOM.

Table 1.4 Samples of input data for different software.

Samples of inputdata			
Cloud userID	**Tenant size**	**Category of software**	**Version of software**
1	60	NS2	Professional
2	20	MATLAB	Standard
3	30	Oracle	Enterprise

Unknown Layer

Each unknown layer node represents a bell-shaped radial basis function that is positioned on a vector in the feature space. From the side-to-side linear combination of nonlinear basis functions in this layer, the RBNNOM can achieve better performance and come close to nonlinear relationships. Some typically used nonlinear functions include the spline function, Gaussian basis function, multiquadratic function, and so on. The Gaussian basis function is applied in this chapter to realize the nonlinear fitting process, which is shown in Equation 1.1:

$$O_{\mathrm{RBF}} = |I_V - C_V|\,\hat{}\,2, \tag{1.1}$$

where I_V is the input vector, C_V is the center vector, and O_{RBF} is the RBF output. Equation 1.1 represents the RBF output, which is equal to square of the difference between the input vector and center vector.

Instruction layer: Each node of this layer is an instruction node that represents one suitable point. The results of the unknown layer nodes are considered by the weights on the lines, and the weighted sum is calculated at the *n*th output node using Equation 1.2:

$$Y_n = \sum\nolimits_{m=1}^{k} A_{nm}\, O_u, \tag{1.2}$$

where A_{nm} links weight, which is measured from unknown layer to production layer, and O_u is the value of the unknown layer [36].

Production layer: The links of the production layer are familiar in reply to some control conditions. ωk represents the output action of the kth rule. The output vector representing the resource characteristics is recorded in Table 1.2, which lists the disk memory size, CPU size, response time, and bandwidth.

Depending on the activation levels of the instructions, the numeric output is given by Equation 1.3:

$$O(Y) = \sum\nolimits_{n=0}^{k} \sum\nolimits_{m=0}^{k} A_{nm} \exp(-|I_V - C_V|\,\hat{}\,2) + bi. \tag{1.3}$$

The locations of the unknown units and the weighted vectors of RBNNOM are found via offline approaches with the training data composed from earlier tasks as presented in Table 1.5. A data preprocessing technique is required before the NN training process. The function of this data preprocessing module is to identify and process abnormal data from

Table 1.5 RBNNOM output for some input data.

Examples of input data			
Disk memory size (GB)	**CPU size**	**Response time**	**Bandwidth**
250	2	3.50	2.50
360	3	3.29	2.30
500	4	5.30	4.80

the samples as well as to quantify the criteria of resource properties. All the earlier four attribute values of input variables and the predictive goal values are normalized into values in the range 0–1. After the network, weight vectors are initialized, and the output of each layer is then figured out using Equations 1.2 and 1.3.

1.3.4 Common Description

Information about Resources

$P = \{N1, N2, \dots Nk\}$ denotes a set of nodes with k representing the total number of existing nodes. The attributes of each node are as follows:

- *Size of memory* ($memRi$, $i = 1, 2, \dots k$): It is defined as the physical size of the memory space allotted to the node with the units in terabytes (TB) or gigabytes (GB).
- *Initiation time* ($inTi$, $i = 1, 2, \dots k$): It is defined as the time taken to initialize a service, which contains VM beginning time and application installation time.
- *Node price* ($cstUi$, $i = 1, 2, \dots k$): It includes the power cost, software cost, and equipment cost. The higher the node price, the higher the quality of the resource concerned.
- *Capacity of CPU* ($cpRi$, $i = 1, 2, \dots k$): It is defined as the capacity and speed of a processor and how many operations it can carry out in a given amount of time, normally referred to using the units *megahertz* (MHz) or *gigahertz* (GHz).
- *Bandwidth* ($bnRi$, $i = 1, 2, \dots k$): It is defined as the rate of data throughput or transfer, measured in bits per second (bit/s). This factor can reflect the speed of interactions among tenants.
- *Expected value of time delay* ($delayTi$, $i = 1, 2, \dots k$): It is the average value of the node's time delay, which generally has a normal distribution.

Tenant Invitation Information

$R = \{R1, R2, \dots Rm\}$ represents a set of requests. In addition to $\{capTj, memTj, banTj, j = 1, 2, \dots m\}$, a request also has the following attributes:

- Request level ($levelj, j = 1, 2, \dots m$): It refers to the priority of the tenants, which is defined in the user database.
- Response time ($resTj, j = 1, 2, \dots m$): It represents the acceptable time taken by the provider to process a particular customer request.
- Penalty for delay ($cosPj, j = 1, 2, \dots m$): It represents the response time taken by the provider that is greater than what is specified in the tenant SLA. Due to the SLA violation, the cloud provider is required to pay a penalty, and this is calculated by using Equation 1.4:

$$CosPj = \begin{cases} A1, & 0 < delayTi < t1 \\ A1 + Ej \times delayTi, & t1 < delayTi \\ 0 & delayTi < 0 \end{cases}, \tag{1.4}$$

where $Ej = \{1, 1.3, 1.5, 1.7, 2, 2.3, 2.5, 2.7, 3\}$. It represents the priority of the loaders.

1.4 Result and Discussion

Figure 1.3 shows that the RBNNOM model provides better iteration as compared to the multivariable control theory model.

Figure 1.4 shows that the RBNNOM model provides better resource allocation among the cloud user, cloud service provider, and tenant as compared to the multivariable control theory model.

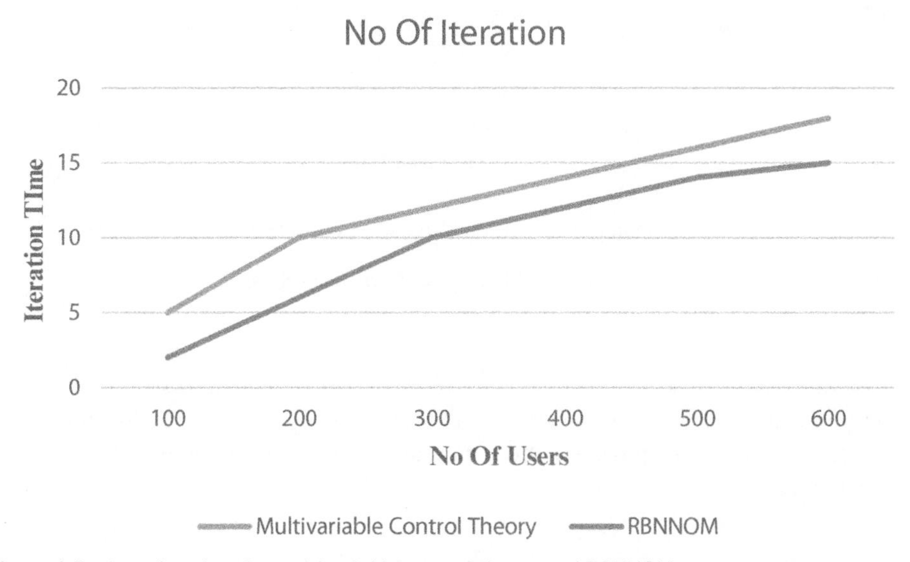

Figure 1.3 Iteration time for multivariable control theory and RBNNOM.

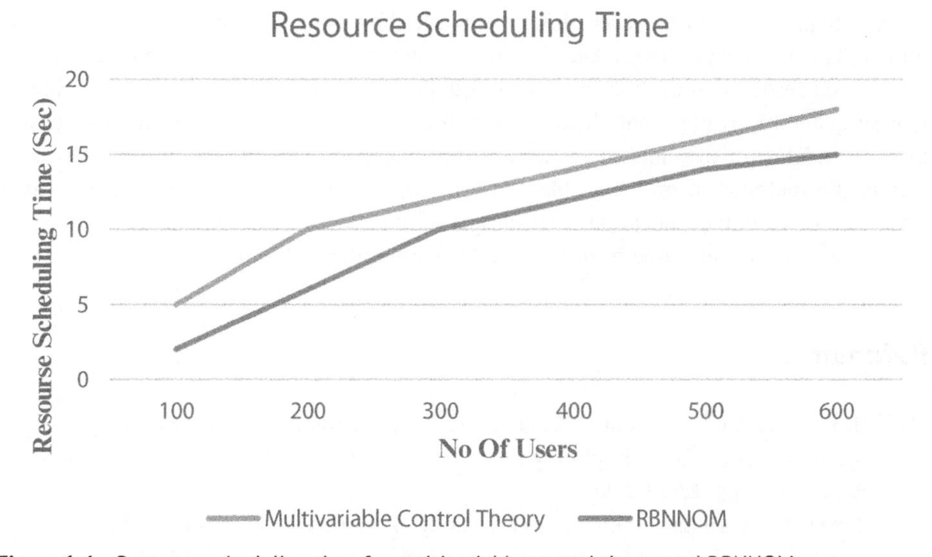

Figure 1.4 Resource scheduling time for multivariable control theory and RBNNOM.

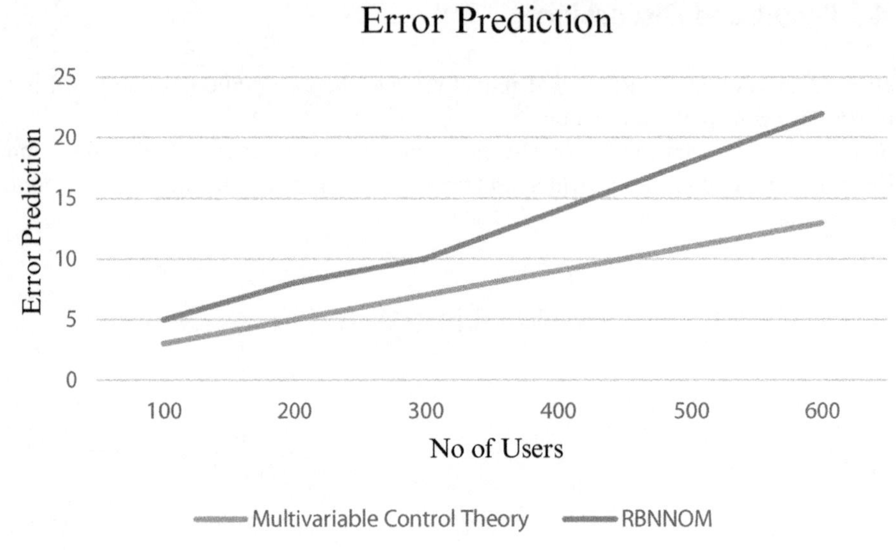

Figure 1.5 Error prediction for multivariable control theory and RBNNOM.

Figure 1.5 shows that the RBNNOM model provides better error prediction among the cloud user, cloud service provider, and tenant as compared to the multivariable control theory model.

1.5 Conclusion

In this chapter, a better resource allocation method by using a multilayer neural network named RBNNOM is proposed. Additionally, the novel method of RBNNOM considers priority-based resource allocation in terms of qualities and quantities and is responsible for QoS with good learning capabilities. Finally, this proposed work analyzes node price and priority weights for the cloud user and cloud provider in the cloud environment by comparing the methods of multivariable control theory and RBNNOM. Also, some critical problems in resource allocation, including its huge scale with complexity constraints, performance of the allocated resource, and QoS, are improved in this study.

References

[1] M. Yu, Y. Yi, J. Rexford, and M. Chiang, "Rethinking virtual network embedding: Substrate support for path splitting and migration," *ACM SIGCOMM Comp. Comm. Rev.*, vol. 38, no. 2, pp. 17–29, 2008.

[2] G. Wei, A.V. Vasilakos, Y. Zheng, and N. Xiong, "A game-theoretic method of fair resource allocation for cloud computing services," *J. Supercomp.*, vol. 54, no. 2, pp. 252–269, 2010.

[3] J. Yang, J. Qiu, and Y. Li, "A profile-based approach to just-in-time scalability for cloud applications." In *2009 IEEE International Conference on Cloud Computing*, IEEE, September 2009; pp. 9–16.

[4] X. Chen, J. Zhang, J. Li, and X. Li, "Resource virtualization methodology for on-demand allocation in cloud computing systems," *Serv. Orient. Comp. Appl.*, vol. 7, no. 2, pp. 77–100, 2013.

[5] L. Ramachandran, N.C. Narendra, and K. Ponnalagu, "Dynamic provisioning in multi-tenant service clouds," *Serv. Orient. Comp. Appl.*, vol. 6, no. 4, pp. 283–302, 2012.

[6] H. Wang, A.L. Johnson, and R.H. Bracewell, "The retrieval of structured design rationale for the re-use of design knowledge with an integrated representation," *Adv. Eng. Info.*, vol. 26, no. 2, pp. 251–266, 2012.

[7] F. Farahnakian, A. Ashraf, T. Pahikkala, P. Liljeberg, J. Plosila, I. Porres, and H. Tenhunen, "Using ant colony system to consolidate VMs for green cloud computing," *IEEE Trans. Serv. Comp.*, vol. 8, no. 2, pp. 187–198, 2014.

[8] K.M. Sim, "Agent-based cloud computing," *IEEE Trans. Serv. Comp.*, vol. 5, no. 4, pp. 564–577, 2011.

[9] C. Papagianni, A. Leivadeas, S. Papavassiliou, V. Maglaris, C. Cervello-Pastor, and A. Monje, "On the optimal allocation of virtual resources in cloud computing networks," *IEEE Trans. Comp.*, vol. 62, no. 6, pp. 1060–1071, 2013.

[10] L. Wu, S.K. Garg, S. Versteeg, and R. Buyya, "SLA-based resource provisioning for hosted software-as-a-service applications in cloud computing environments," *IEEE Trans. Serv. Comp.*, vol. 7, no. 3, pp. 465–485, 2013.

[11] F. Tao, Y. Feng, L. Zhang, and T.W. Liao, "CLPS-GA: A case library and Pareto solution-based hybrid genetic algorithm for energy-aware cloud service scheduling," *Appl. Soft Comp.*, vol. 19, pp. 264–279, 2014.

[12] G. Peng, H. Wang, J. Dong, and H. Zhang, "Knowledge-based resource allocation for collaborative simulation development in a multi-tenant cloud computing environment," *IEEE Trans. Serv. Comp.*, vol. 11, no. 2, pp. 306–317, 2016.

[13] Z. Xiao, W. Song, and Q. Chen. "Dynamic resource allocation using virtual machines for cloud computing environment," *IEEE Trans. Parallel Distrib. Sys.*, vol. 24, no. 6, pp. 1107–1117, 2013.

[14] L. Wang, F. Zhang, J.A. Aroca, A.V. Vasilakos, K. Zheng, C. Hou, D. Li, and Z. Liu, "GreenDCN: A general framework for achieving energy efficiency in data center networks," *IEEE J. Select. Areas Comm.*, vol. 32, no. 1, pp. 4–15, 2013.

[15] A.G. Kumbhare, Y. Simmhan, M. Frincu, and V.K. Prasanna, "Reactive resource provisioning heuristics for dynamic dataflows on cloud infrastructure," *IEEE Trans. Cloud Comp.*, vol. 3, no. 2, pp. 105–118, 2015.

[16] L. Mashayekhy, M.M. Nejad, D. Grosu, and A.V. Vasilakos, "An online mechanism for resource allocation and pricing in clouds," *IEEE Trans. Comp.*, vol. 65, no. 4, pp. 1172–1184, 2015.

[17] H. Zhang, H. Jiang, B. Li, F. Liu, A.V. Vasilakos, and J. Liu, "A framework for truthful online auctions in cloud computing with heterogeneous user demands," *IEEE Trans. Comp.*, vol. 65, no. 3, pp. 805–818, 2016.

[18] G. Zhang, X. Zhu, W. Bao, H. Yan, and D. Tan, "Local storage-based consolidation with resource demand prediction and live migration in clouds," *IEEE Access*, vol. 6, pp. 26854–26865, 2018.

[19] R.N. Calheiros, E. Masoumi, R. Ranjan, and R. Buyya, “Workload prediction using ARIMA model and its impact on cloud applications’ QoS,” *IEEE Trans. Cloud Comp.*, vol. 3, no. 4, pp. 449–458, 2014.

[20] H. Morshedlou and M.R. Meybodi, “Decreasing impact of sla violations: A proactive resource allocation approachfor cloud computing environments,” *IEEE Trans. Cloud Comp.*, vol. 2, no. 2, pp. 156–167, 2014.

[21] Z. Yan, X. Li, M. Wang, and A.V. Vasilakos, “Flexible data access control based on trust and reputation in cloud computing,” *IEEE Trans. Cloud Comp.*, vol. 5, no. 3, pp. 485–498, 2015.

[22] M. Ali, R. Dhamotharan, E. Khan, S.U. Khan, A.V. Vasilakos, K. Li, and A.Y. Zomaya, “SeDaSC: Secure data sharing in clouds,” *IEEE Sys. J.*, vol. 11, no. 2, pp. 395–404, 2015.

[23] R. Tolosana-Calasanz, J. Diaz-Montes, O.F. Rana, and M. Parashar, “Feedback-control & queueing theory-based resource management for streaming applications,” *IEEE Trans. Parallel Distrib. Sys.*, vol. 28, no. 4, pp. 1061–1075, 2016.

[24] K. Gai, L. Qiu, H. Zhao and M. Qiu, "Cost-Aware Multimedia Data Allocation for Heterogeneous Memory Using Genetic Algorithm in Cloud Computing," in *IEEE Transactions on Cloud Computing*, vol. 8, no. 4, pp. 1212–1222, 1 Oct.–Dec. 2020, doi: 10.1109/TCC.2016.2594172.

[25] J. Rao, Y. Wei, J. Gong, and C.Z. Xu, “QoS guarantees and service differentiation for dynamic cloud applications,” *IEEE Trans. Netw. Serv. Manag.*, vol. 10, no. 1, pp. 43–55, 2013.

[26] M.A. Islam, S. Ren, G. Quan, M.Z. Shakir, and A.V. Vasilakos, “Water-constrained geographic load balancing in data centers,” *IEEE Trans. Cloud Comp.*, vol. 5, no. 2, pp. 208–220, 2015.

[27] S. Gong, B. Yin, W. Zhu, and K. Cai, “An adaptive control strategy for resource allocation of service-based systems in cloud environment.” In *2015 IEEE International Conference on Software Quality, Reliability and Security-Companion*, IEEE, August 2015; pp. 32–39.

[28] Z. Song, Y. Sun, J. Wan, and P. Liang, “Data quality management for service-oriented manufacturing cyber-physical systems,” *Comp. Electr. Eng.*, vol. 64, pp. 34–44, 2017.

[29] L. Baresi, S. Guinea, A. Leva, and G. Quattrocchi, “A discrete-time feedback controller for containerized cloud applications.” In *Proceedings of the 2016 24th ACM SIGSOFT International Symposium on Foundations of Software Engineering*, November 2016; pp. 217–228.

[30] Y. Xia, M. Zhou, X. Luo, Q. Zhu, J. Li, and Y. Huang, “Stochastic modeling and quality evaluation of infrastructure-as-a-service clouds,” *IEEE Trans. Autom. Sci. Eng.*, vol. 12, no. 1, pp. 162–170, 2013.

[31] S. Farokhi, E.B. Lakew, C. Klein, I. Brandic, and E. Elmroth, “Coordinating cpu and memory elasticity controllers to meet service response time constraints.” In *2015 International Conference on Cloud and Autonomic Computing*, IEEE, September 2015; pp. 69–80.

[32] P.S. Saikrishna, R. Pasumarthy, and N.P. Bhatt, “Identification and multivariable gain-scheduling control for cloud computing systems,” *IEEE Trans. Control Sys. Technol.*, vol. 25, no. 3, pp. 792–807, 2016.

[33] S. Gong, B. Yin, W. Zhu, and K.Y. Cai, “Adaptive resource allocation of multiple servers for service-based systems in cloud computing.” In *2017 IEEE 41st Annual Computer Software and Applications Conference (COMPSAC)*, IEEE, July 2017; vol. 2, pp. 603–608.

[34] S. Gong, B. Yin, Z. Zheng, and K.Y. Cai, "Adaptive multivariable control for multiple resource allocation of service-based systems in cloud computing," *IEEE Access*, vol. 7, pp. 13817–13831, 2019.

[35] T. Senthil Murugan and N. Vijayaraj, "Skewness based dynamic resource allocation in cloud using heterogeneous," *Int. J. Innovative Tech. Expl. Eng.*, vol. 8, no. 7, pp. 1449–1455, May 2019, ISSN: 2278–3075 (Online).

[36] N. Vijayaraj' and T. Senthil Murugan, "Fair resource allocation in cloud using control theory model" has been accepted and considered for publication in the Journal of Advanced Research.

2

Internet of Things for Human-Activity Recognition Based on Wearable Sensor Data

Dr. Vikram Rajpoot[1], Sudeep Ray Gaur[2], Aditya Patel[3], and Dr. Akash Saxena[4]

[1] *GLA University, Mathura*
[2,3] *LNCT College, Bhopal*
[4] *CITM, Jaipur*

2.1 Introduction

2.1.1 Internet of Things

What is the *Internet of things* (IoT)? In fact, there is no absolute or official definition for *IoT*. It is basically the body of practical knowledge that is used to change the world into a better place.

Kevin Ashton, also known as the inventor of IoT, first coined the term "Internet of things" in 1999. He used the term to describe a system where anyone can get connected to the virtual world through the Internet [5].

The term *Internet of things*, generally referred as *IoT*, is a concept where computing potential is provided to objects, sensors, and everyday items not normally considered computers, allowing these devices to bring about, exchange, and consume data with minimal human intervention [5].

In today's world, we can see rapid changes in actual usages and utilities of things. There are many examples which can be put forward to show the cause of this alteration.

Advances in wearable electronics have the potential to affect a wide range of health applications. For example, diagnosis and follow-up for many health problems, such as motion disorders, depend currently on the behavior observed in a clinical environment. Specialists analyze gait and motor functions of patients in a clinic and prescribe a therapy accordingly. As soon as the person leaves the clinic, there is no way to continuously monitor the patient and report potential problems. Another high-impact application area is obesity-related diseases, which claim about 2.8 million lives every year. Automated tracking of physical activities of overweight patients, such as walking, offers tremendous value to health specialists, since self-recording is inconvenient and unreliable. As a result, HAR, with low-power wearable devices, can revolutionize health- and activity-monitoring applications [5]. There has been growing interest in HAR with the prevalence of low-cost motion sensors and smartphones. For example, accelerometers in smartphones are used to recognize activities such as sit-to-stand exercise, lying down, walking, and jogging. This information is used for rehabilitation instruction, fall

Sensor Data Analysis and Management: The Role of Deep Learning, First Edition. Edited by A. Suresh, R. Udendhran, and M.S. Irfan Ahmed.
© 2021 John Wiley & Sons, Ltd. Published 2021 by John Wiley & Sons, Ltd.

detection of the elderly, and reminding users to be active [6]. Furthermore, activity tracking also facilitates physical activity, which improves the wellness and health of its users. HAR techniques can be broadly classed based on when training and activity recognition take place. One of the previous studies collected sensor data before processing [7]. Then, both classifier design and activity recognition were performed. Hence, they had limited applicability. A more recent study trained a classifier, but processed the sensor data online to recognize the activity. However, to date, there is no technique that can perform both online training and activity recognition. Online training is crucial, since it needs to adapt to new and a potentially large number of users who are not involved in the training process. To this end, this chapter presents the first HAR technique that continues to train online to adapt to its user. A vast majority, if not all, of recent HAR techniques are deployed in smartphones. The major motivations behind this choice are their widespread use and easy access to integrated accelerometer and gyroscope sensors. We argue that smartphones are not suitable for HAR for three reasons. First, patients cannot always carry a phone as prescribed by the doctor. Even when they have the phone, it is not always in the same position (e.g., in the hand or in the pocket), which is typically required in these studies. Second, mobile operating systems are not designed for meeting real-time constraints. Due to the same reason, the power consumption is in the order of watts (more than 100× of our result). Finally, research is limited to sensors integrated in the phones, which are not specifically designed for HAR [8].

2.1.2 IoT and the Smart Environment

After the concept of automation emerged, almost everything around us—from a door in the house to the operation of heavy machinery—is moving toward becoming automated.

In simpler words, it is basically the forging of a "smart environment" in which every possible thing could be controlled using computers, whose computing potential is much more than that of an average person [8]. In a smart environment, all things are linked with each other, and they can be controlled easily using this network. A smart environment includes smart homes, smart cities, smart manufacturing, etc.

A smart environment is basically an addendum of persuasive or ubiquitous computing. It assists the idea of a world that is managed by the use of sensors. These sensors and computer devices are integrated with everyday objects and things.

IoT is a new conceptualization in which all sensing objects can be connected to the Internet to have remote and persistent gain or access to their data. This gain or access allows for taking action in a much faster way, with great results and much more data involved. The IoT can be summed up as the use of sensors. By using sensors, the data are made available to create the required datasets. The sensors which can be used are temperature sensor, location-tracker sensor, weather-related factors sensor, stock indicator, alert sensor, and other industry-associated variables. In our case, we need to focus on the data related to daily human activities such as sitting, walking, limping, etc. In short, any sensor that can be connected to a computational device and to the Internet is part of an IoT prototype [9].

Many IoT-based devices are available in the market nowadays. These devices consist of the sensors through which data can be made available and therefore get accessed. There are a whole lot of IoT devices that are now created for consumer use, including connected vehicles, home automation equipment, wearable technology, connected health monitors, and appliances with remote monitoring capabilities.

Similarly, using light sensors, we can make a device that basically checks for the intensity of light, whether it is more than a threshold or not [9]. This can be used in monitoring crops, which can then be deployed in the field of farming and agriculture.

You may have also heard of smart traffic-light management. In this, there is utilization of a sound sensor to check if there is more honking than the threshold value; depending on this, the timer at the signal resets to the maximum and people wait for more time. This is an interesting example, since this system gauges people's actions and utilizes the results to help enforce the law and lessen noise pollution. This indicates the vastness of IoT's scope of use, and points to how everything around us can be transformed using this concept, leading to an automated world [9].

HAR has attracted significant research interest due to its applications in health monitoring and patient rehabilitation. Recent research on HAR focuses on using smartphones, owing to their widespread use. However, this leads to inconveniences, issues arising from the limited choice of sensors, and the inefficient use of resources, since smartphones are not designed for HAR.

2.1.3 HAR

We are aware of how IoT is being implemented in various fields—be it for controlling a light bulb or for managing human health in the hospital, many aspects of life can now be controlled through the use of IoT. Human health can be monitored using several daily-use devices such as fitness bands, and patients can therefore be made aware if any mishap is about to take place [10].

Similarly, this classical application of ubiquitous or persuasive computing can now be upgraded by a few steps to an IoT proposal for an activity-recognition application: HAR. These applications have been explored, assessed, and developed to the point that many products commonly available in the market now have inbuilt HAR systems [10]. This can be seen very commonly in the mobile applications of these products, which nowadays come with inbuilt HAR systems.

Some of the devices in which the HAR system can be easily seen are fitness bands, smart watches, smart glasses, and smart shoes. These devices help in keeping track of not only someone's day-to-day activities but also that person's sleep to see whether there is any kind of variation in the data present or if it is all standard.

The HAR system consists of observing a person's positioning and movement, then isolating some features, classifying these from the full range of activities considered, and deciding which activity was performed in a standard way [10]. This is done by matching the current activity data, extracted by computing sensor inputs, against a preselected activity or data list that contains the data of activities that are already classified or categorized in a list of activities (a dataset).

HAR systems have several methods to perform recognition, one being an artificial vision-assisted system [1]. The main characteristics of this system are controlled stability–plasticity conduct, controlled reliability threshold, both learning and self-evaluation of the output reliability, and high reliability based on high-level multiple feedback. The system has been designed using a commutable perspective. And that is how the system is made potentially suitable in a large variety of vision applications.

Using artificial vision-assisted recognition system, we can train our model or device just with algorithms and machine learning using object detection, etc., which we won't discuss here, as we will mainly focus on HAR using IoT devices [10]. But still, this former system discussed has shown great progress in performance on real-world data and complete recognition of the objects or images used for testing.

Regardless of the many benefits and approval, it also has some disadvantages, such as accuracy and coverage, and it is also very costly to operate on. Therefore, we have to look for some other replacement that can remove the drawbacks present in artificial vision-assisted recognition systems.

Hence, another method that can overcome the challenges faced in artificial vision-assisted system involves using on-body sensor systems or HAR-assisted wearables. Now let us focus on the main topic of this chapter, that is, IoT-based HAR systems, or simply IoT-HAR systems [11].

In this approach, we basically rely on wearables consisting of sensors that look throughout the body of the test subject, helping to recognize the activity performed. This method can be successfully implemented by requiring the subject to wear one or more devices for pre-established periods of time. Some systems even require that the device(s) be permanently worn by the person. These devices could be those within another device, such as a smartphone [11].

Our smartphones or mobile phones also consist of sensors through which basic assessing can be done successfully. For example, you may have noticed how sensors in your smartphones work. While talking on the phone, when we place the phone close to our ear or any part of the skin, the screen turns off. This happens because of the proximity sensor, which takes notice if some changes take place.

Many other sensors too are present in mobile phones, such as accelerometer, gyroscope, magnetometer, etc. These sensors can be used to create records of the day-to-day activities performed in a person's life [11]. When tracking these activities, data are produced using algorithms, and this data are then classified such that it can be distinguished whether the activity is already known to our dataset, or if it is new, so that any new data can be added to the existing dataset, modifying it as required.

This system is very helpful to patients suffering from chronic diseases like heart diseases or patients with critical or noncritical health conditions who need to see their own performance and keep track of their daily activities. In simpler words, it means that this system can be used by people who are health conscious and who care about themselves.

The system mentioned in the preceding text has been made ubiquitously available with the widespread use of mobile devices. But what if a person does not have a mobile phone, or does not want to retain a mobile phone while jogging or exercising? A solution is available for that too [11].

Wearables are devices that a person can wear while continuing with their daily activities. These devices are now being used a lot. Every two out of ten people have fitness bands or smart watches, but they are a bit costly. Wearables can help in keeping a record of a person's activities, and if there is some unusual activity, the person gets a warning on their mobile screen. Therefore, these devices help people in maintaining their daily activities without hindrance.

2.2 System Architecture

HAR has a problem in foreseeing what a person is doing based on tracing their movement using sensors. Movements are often normal indoor activities such as standing, sitting, jumping, going upstairs, etc., judged using the sensors [12].

Sensor vests are also present in the market today to be positioned against the chest of a person. The sensors are often located on the subject via a smartphone or a vest and often record accelerometer data in three dimensions (x, y, and z). Here, the basic ideology is that, once the activity is recognized and known, the intelligent system can offer some assistance.

Before this intelligent system can offer assistance, it is mandatory to have some data made available by the help of the sensors. There are many kinds of wearables available that can be linked with mobile phones by using some applications. Health applications present in mobile phones can keep track of activities such as:

- Working at computer (or simply working)
- Standing
- Walking
- Going upstairs/downstairs
- Talking while walking
- Talking while standing
- Sleeping
- Running/jogging, etc.

After getting or tracing the activity performed by a person, the intelligent HAR system can actually recognize the activity [12].

In HAR systems, as with wearables, all activities are assessed and therefore categorized into accustomed and unaccustomed activities. In other words, it is simply checked whether the activity performed by the human is already known or new to the system.

This system is often implemented on mobile or wearable devices. The technique implemented in this system basically recognizes activities by using an algorithm [12]. This happens in two phases: evaluation phase and validation phase, as presented in Figure 2.1.

2.3 Evaluation Phase

In the evaluation phase, we have a group of metrics to quantify the activity-recognition performance, such as accuracy, consistency, etc., and there should be straightforward and good-level complexity in order to generate the data. Here, by using the precision and accuracy of the sensors present in the device, we are ready to evaluate activity recognition. In other words, during this phase, we develop the dataset consistent with the data assessed by

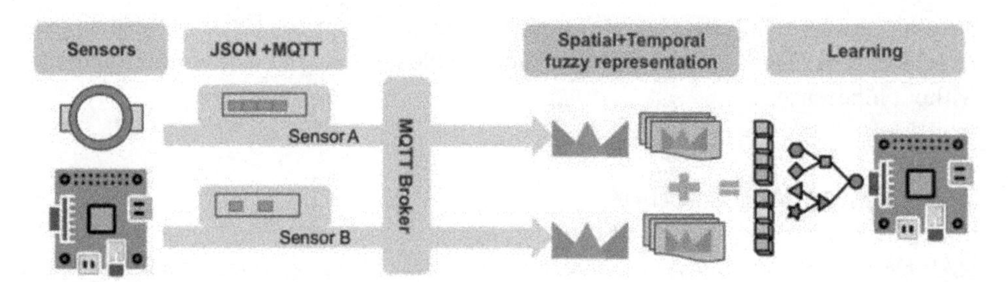

Figure 2.1 Sequence of sensors and its workflow.

the wearable devices using an IOT-based HAR system [13]. For this, there are some steps that are followed in evaluating an activity. They are as follows:

1. *Loading of data/data collection*: This step consists of loading the data into the system for evaluation from all the sensors present in the device used. The data are captured according to the type of sensor used. In other words, it means that the data made available for the system are dependent on the sensor used. In this phase, all the activities are considered, whether known or unknown; we do not have to think about it, as we do not have to classify activities in this phase. In the next step, we also keep track of the activities performed. It means that we record information about the activity performed by the test subject [2]. This log consists of information such as time, duration, etc., so that it is easy to use the data in the future.
2. *Selection/feature extraction*: During the previous step, data collection, there could be a possibility that the sensor present in the device is not compatible with the system and collects wrong data. Hence, this step plays a very important role in getting limited but accurate data. The name of this step already suggests how data selection is done in the device. During this selection of data, many structural and statistical features are applied to the data. Structural features try to find out the association between the signals. The statistical feature performs a mutation on the signal using statistical information or the existing data. Some of the features of the signals could be the mean, standard deviation, etc. The most common mutations, variations, or transformations performed are the Fourier and wavelet transforms [2]. During this step, signal processing is done by using structural and statistical features to abolish or get rid of the noise and other kinds of disturbances that may be present in the data while loading into the system for analysis.
3. *Comprehending the activity using data/learning*: The comprehending or learning is the final step of this phase in which development of a recognition model is done. Various algorithms are used to develop the recognition model. Some of them are iteration algorithms, domain transformation, fuzzy logics, Markov model, regression model, etc. Therefore, in this phase, data are extracted from device sensors, features are selected, and then the model is developed. Learning and developing models are the main parts of this phase [14].

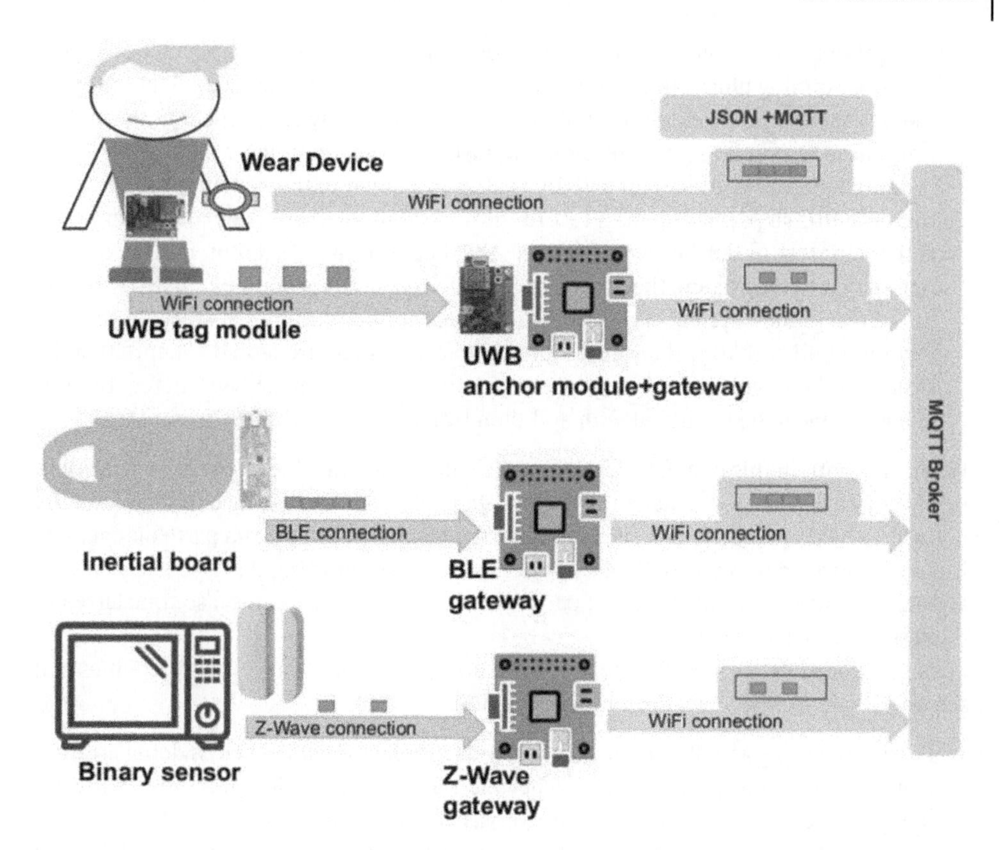

2.4 Validation Phase

This is the final phase of the HAR system. In this phase, recognition of the activity is done, which is performed as a result of the previous phase, the evaluation phase.

In the validation phase, a crucial step in recognizing tasks is to separate the available data into training and testing sets. So, the data made available to the system in the previous phase are then used and categorized into known or unknown. It consists of a classifier that helps in classifying the data accordingly. The following are some steps that are followed for validating an activity:

1. *Loading of data/data collection*: This step consists of loading the sensor data into the system for evaluation. The data are captured according to the type of sensor used [13]. As in the evaluation phase, there is no need of having preliminary knowledge about the activity, and therefore there is no need of keeping track of the activities.
2. *Selection/feature extraction*. This step is identical to the evaluation phase in order to get rid of any kind of inappropriate data present in the dataset, which can cause hindrances in the process of recognition [14].

3. *Recognition*: The recognition is the final step of the validation phase, which can also be called as training phase. In this phase, the less erroneous data made available through the previous steps are recognized by the system. As already mentioned, in the validation phase, a crucial step in recognizing tasks is to separate the available data into training and testing sets. If an activity is already known to the system through recognition in this step, it is kept as a record in the testing set. And if any alteration in the activity present in the dataset is detected, the dataset is modified. But, if the activity is not known to the system, then the system is alerted that this particular activity is new to the recognition system. This data is then kept in the training set, since, in this set, records of all activities that are unknown to the system are stored. Then, after the system is done with the whole process of understanding the activity, that activity is given a name to be identified with and then transferred into the testing set [15].

Hence, in short, in this phase, validation of the data made available by the sensors present in the device is done and matched with the data present in the existing dataset. Matching of the data means the activity is recognized by the testing set. If not, that particular activity is put under the training set of the recognizer by the classifier using classifying algorithms, and then a dataset is created for that particular activity. After this, it is also transferred to the testing set of the activities.

As mentioned in the preceding text, we will now go through some human activities and see how the HAR system works for these activities.

1. *Playing*: Working activity can involve many day-to-day activities such as doing errands, playing games and sports, etc. It was beyond human imagination until now that technology could become so powerful and pervasive that all our activities or the things we do in our daily lives can be detected and tracked by intelligent systems. This has been made possible because we train the technology to track and recognize whatever actions humans are capable of, which now helps us take care of our health. Therefore, for this to work, the system needs to capture each and every moment and generate the associated data [15].

   Here, we take a "vest" as a sensor device. Using this vest can help us not only in noticing the useable things but also in capturing and visualizing parts of the activities. In such a vest, data captured are totally dependent on the type of sensors available in the device. Some vests consist of a movement sensor, and some can also have temperature sensors and heart rate sensors. This depends on the type of vest you are likely to wear [15].

   For capturing day-to-day activities as mentioned in the preceding text, visualization of these activities can help considerably in training the HAR system. The vest we have used in this activity recognition consists of wireless sensors such as GPS tracker, accelerometer, gyroscope, a Raspberry Pi module, and a Pi Camera.

   When the vest is worn, it gets connected to the application, so that the data generated during any activity can be recorded in the cloud. The data are recorded in (x, y, and z) dimensions or coordinates.

   For example, we take the activity of playing football in this case. Here, while running for the ball on the field, the accelerometer and the dimensions captured can help the system to recognize that the player is running, jumping, or engaged in some other activity at any given time [15].

The Raspberry Pi module and the Pi Camera used in the vest capture every moment. Using the captured data and applying object detection and image processing, we can train the system more efficiently to understand the actions of the player.

IoT plays a major role in making this HAR system more enhanced. All the wireless sensors as well as the Pi modules used here offer a practical approach to implementing the IoT concept.

2. *Walking*: The movement of the human body on a plane surface is termed *walking*. In the context of activity recognition, it means detecting when the associated movement is made by a human body and how much distance it has covered.

   Let us consider a smart watch or a fitness band, which is a device to wear on the wrist while we go about our daily lives. These bands and watches are now empowered with many advanced technological features, and they contain activity trackers that sense the movements made by humans using a three-axis (x, y, and z) accelerometer.

   The incorporated sensor detects movements and captures data related to walking speeds, the direction in which the wearer is walking, distance walked, etc. All the information regarding this activity is recorded and kept for the system to process [15].
3. *Running*: Similar to the walking action, running too can be recognized using wearables such as fitness bands, smart watches, and other similar devices. The device keeps track of the movements of the body along with some other attributes such as speed, direction, duration, etc.

   There are also a lot of related applications available today. The basic kinds of day-to-day activities can also be tracked by smartphones, which almost everyone uses today.

   Mobile applications make use of the sensors available in mobile handsets, such as accelerometers and gyroscopes, to determine if the user is sitting or running. This is done using the three-axis accelerometer [16].
4. *Sprinting*: This is also similar to running. However, running, jogging, and sprinting are different activities. Sprinting involves running very fast over a short distance. Running is faster than jogging, jogging is faster than walking, and, obviously, walking is faster than sitting [16]. The sensors are used to maintain records of the activity's data, helping the system to evaluate and validate the data.
5. *Talking while sitting*: On the one hand, sitting is a very normal and basic activity performed by all human beings. Wearable devices register this activity when there is no movement of the body along any axis; in short, when the body is at rest.

   Talking, on the other hand, is also one of the basic activities performed by the humans. The way of talking, the tone, the language used, etc., are all attributes that can be considered for training an HAR system in language recognition, and also to integrate a talk-back feature if desired. Hence, talking and movement of the body are both clubbed in this type of activity recognition [16].

   Let us consider a smart watch, for example. Smart watches available in the market today contain a display and a microphone, and several sensors through which a human body's day-to-day activities can be recognized, which in this case will be null. And the microphone is used to listen to the person's speech and then apply filters to whatever is captured by using mathematical and translation tools, such as Google's speech recognizer and many others.

Hence, this recognition can play a major role in enhancing the dataset—and therefore can also help in its implementation on various devices such as smart speakers, etc., in order to assist humans in daily activities [16].

6. *Talking while walking*: This activity recognition is quite similar to the abovementioned activity, that is, talking while sitting, for which we discussed, with the help of an example, how the system is trained using the sensors and microphone available in the device, and how a dataset is then developed.

   Similarly, by using a smart watch, we can help the system to recognize the activities performed by a human being in day-to-day life. Using the three-axis accelerometer in the watch, we can keep track of the human body as it performs movements or moves here and there. Using the microphone will help in generation of the dataset, which would consist of data generated while performing speech recognition.

   These data are very much helpful, as there are many robots and smart devices being manufactured with smart features that help in getting many of our daily chores done.

   Some of these devices are smart speakers, smart watches, smart televisions, smart mirrors, etc. All these devices notice the everyday routine of a human being, from the beginning of the day till the person sleeps. And some devices keep on monitoring the activities of the body even while the person sleeps [15].

   So, focusing on talking while walking, movement is recognized as described in the earlier text, and speech recognition is done using the microphone. In addition to smart watches or smart bands, other devices such as smartphones too can be used, since they too have sensors available such as accelerometers, gyroscopes, microphones, etc.

   Apart from smartphones, smart speakers too are available with inbuilt HAR systems. This has also been proven to be helpful in training the system. However, this is also a possible of threat to people's privacy and security, as they could be inadvertently sharing some private or confidential information that they do not want shared with anyone—for example, you may have heard the news of Google's confession that it keeps track of people's daily activities and listens in to their conversations.

7. *Sleeping*: Sleep is the best way to rid ourselves of strain and stress and to maintain a peaceful, fresh mind. We know that sleep is of utmost importance for every human being. However, people do not get adequate sleep nowadays due to hectic and stressed out lifestyles [15].

   Sleeping activity is recognized by this HAR system, which further helps in maintaining a dataset, so that people can be made aware that they are not getting enough sleep.

   We already know that fitness bands and smart watches come with sleep trackers, which act in a similar way as discussed earlier. Since bands and watches have sensors, they are able to detect whether a person is awake or not. There are many factors that can help the device detect if a person is sleeping or not, such as respiration rate, heart rate, etc. If a person is awake, these factors would register differently, and if a person is highly stressed out, then these factors would deviate considerably from normal or average levels. At this point, the device alerts the person to take care of their sleep.

8. *Going up/down on stairs*: Buildings often have several floors, which we negotiate using stairs or elevators. People often prefer elevators to stairs forgoing up and down, owing to the effort involved in using stairs. However, taking stairs instead of elevators could have some health benefits [16].

This activity recognition is made possible by using mobile phones or any other sensing devices. Let us take a very practical example, iPhones have their own application for health monitoring, which can record all human activities throughout the day, including the number of floors traveled. This is done using the sensors present in these phones. This activity can also be tracked in fitness bands and smart watches.

9. *Studying*: This activity recognition seems rather meaningless, but it is, in fact, quite beneficial and likely to be implemented in future as well. Activities such as the studying done by a student can be recognized by this HAR system so as to train its model and make a dataset. As the activity of doing errands is recognized by the system, this dataset can be used to make automated robots or devices that can perform the entire activity automatically. In a very similar way, studying activity is recognized by this HAR system, but with a slight difference. Many other activities can be performed by automated devices in the future, but studying is an activity that is necessary for everyone and therefore it cannot be done by a device.

   However, in future, there could be devices that can help in tutoring students and children as well as adults if they want to learn. You may have heard the news that an HR person has been replaced by a robot or a screen that can interview many people simultaneously. This technology works efficiently, but is also affecting people and their jobs, as well as their skills [16].

   Tutoring devices could be developed in the future that can put special efforts on a child to focus on and resolve the problems that the child is facing.

   However, for this to be implemented, the basic need is to prepare a dataset that can be made only by recognizing the activity using sensing devices (see Tables 2.1 and 2.2).

Smart phones are all pervasive and becoming more and more innovative. This has been changing the daily lives of individuals and has opened the door for fascinating information-mining applications. Human-action recognition is the center structure hidden behind these applications. It takes the crude sensor's output as sources of data and predicts a user's movements and actions [17]. Various classifiers are utilized for assessing the recognized data. The strategy of utilizing normal probabilities was found to be the most effective for movement recognition, outperforming all other classifiers. This chapter further demonstrated that the recognition strategy can identify exercises irrespective of the cell phone's position. An area for potential research would be to find novel methods of movement

Table 2.1 Sequence activities of some case scenes.

Scene 1	Sleeping Using toilet Preparing lunch Eating Watching TV Using phone→Dressing→Brushing teeth→Exiting
Scene 2	Entering→Drinking→Using toilet→Using phone→Exiting
Scene 3	Entering→Drinking→Using toilet→Dressing→Cooking→Eating→Sleeping
Scene 4	Entering→Using toilet→Dressing→Watching TV→Cooking→Eating→Brushing teeth→Sleeping
Scene 5	Entering→Drinking→Using toilet→Dressing→Cooking→Eating→Using phone→Brushing teeth→Sleeping

Table 2.2 Frequency (number of steps) for each activity and scene.

Activity	Scene 1	Scene 2	Scene 3	Scene 4	Scene 5
Sleeping	10	0	16	11	11
Using toilet	13	10	6	10	14
Cooking	29	0	27	17	24
Eating	36	0	40	46	41
Watching TV	20	0	0	16	0
Using phone	14	17	0	0	15
Dressing	15	0	21	16	17
Brushing teeth	21	0	0	18	18
Exiting	3	2	0	0	0
Entering	0	3	4	2	3
Drinking	0	16	7	0	12

recognition. Future research could also concentrate on (1) recognizing new exercise routines [17]; (2) trying to gather information from more clients of different ages; and (3) removing highlights that could fine-tune the segregation of various exercises.

References

[1] J. Gubbi, R. Buyya, S. Marusic, and M. Palaniswami, "Internet of Things (IoT): A vision, architectural elements, and future directions," *Future Gener. Comput. Syst.*, vol. 29, pp. 1645–1660, 2013.

[2] G. Roncancio, M. Espinosa, and M.R. Perez, "Spectral sensing method in the radio cognitive context for IoT applications," In *Proceedings of the 10th IEEE International Conference on Internet of Things (iThings 2017)*, Exeter, Devon, UK, 21–23 June 2017; pp. 1–6.

[3] J. Cabra, D. Castro, J. Colorado, D. Mendez, and L. Trujillo, "An IoT approach for wireless sensor networks applied to E-health environmental monitoring," In *Proceedings of the 10th IEEE International Conference on Internet of Things (iThings 2017)*, Exeter, Devon, UK, 21–23 June 2017; pp. 14–22.

[4] N. Velasquez, C. Medina, D. Castro, J.C. Acosta, and D. Mendez, "Design and development of an IoT system prototype for outdoor tracking," In *Proceedings of the International Conference on Future Networks and Distributed Systems – ICFNDS'17*, Cambridge, UK, 19–20 July 2017; pp. 1–6.

[5] M. Teran, J. Aranda, H. Carrillo, D. Mendez, and C. Parra, "IoT-based system for indoor location using bluetooth low energy," In *Proceedings of the IEEE Colombian Conference on Communications and Computing (COLCOM2017)*, IEEE Xplore Digital Library, Cartagena, Colombia, 16–18 August 2017.

[6] F. Ganz, P. Barnaghi, and F. Carrez, "Information abstraction for heterogeneous real world Internet data," *IEEE Sens. J.*, vol. 13, pp. 3793–3805, 2013.
[7] C. Perera, C.H.I.H. Liu, S. Jayawardena, and M.A. Chen, "Survey on Internet of things from industrial market perspective," *IEEE Access*, vol. 2, pp. 1660–1679, 2014.
[8] Fitbit, "Heart rate tracker: Fitbit charge 2TM." Available online: https://misfit.com/fitness-trackers (accessed on 25 November 2017).
[9] Misfit, "Misfit: Fitness trackers & wearable technology—Misfit.com." Available online: https://www.fitbit.com/home (accessed on 25 November 2017).
[10] C.H. Liu, "A survey of context-aware middleware designs for human activity recognition communications," *IEEE Commun. Mag.*, vol. 52, pp. 24–31, 2014.
[11] F. Sikder and D. Sarkar, "Log-sum distance measures and its application to human-activity monitoring and recognition using data from motion sensors," *IEEE Sens. J.*, vol. 17, pp. 4520–4533, 2017.
[12] Z. Wang, D. Wu, J. Chen, A. Ghoneim, and M.A. Hossain, "Human activity recognition via game-theory-based feature selection," *IEEE Sens. J.*, vol. 16, pp. 3198–3207, 2016.
[13] A. Testoni and M. Di Felice, "A software architecture for generic human activity recognition from smartphone sensor data," In *Proceedings of the 2017 IEEE International Workshop on Measurement and Networking (M&N)*, Naples, Italy, 27–29 September 2017; pp. 1–6.
[14] R. Poppe, "A survey on vision-based human action recognition," *Image Vis. Comput.*, vol. 28, pp. 976–990, 2010.
[15] L. Mo, F. Li, Y. Zhu, and A. Huang, "Human physical activity recognition based on computer vision with deep learning model," In *Proceedings of the IEEE Instrumentation and Measurement Technology Conference*, Taipei, Taiwan, 23–26 May 2016.
[16] B. Boufama, "Trajectory-based human activity recognition from videos," In *Proceedings of the 3rd International Conference on Advanced Technologies for Signal and Image Processing—ATSIP'2017*, Fez, Morocco, 22–24 May 2017; pp. 1–5.
[17] G. Chetty and M. White, "Body sensor networks for human activity recognition," In *Proceedings of the 2016 3rd International Conference on Signal Processing and Integrated Networks (SPIN)*, Noida, India, 11–12 February 2016; pp. 660–665.

3

Evaluation of Feature Selection Techniques in Intrusion Detection Systems Using Machine Learning Models in Wireless Ad Hoc Networks

T.J. Nagalakshmi[1], *M. Balasaraswathi*[2], *V. Sivasankaran*[3], *D. Ravikumar*[4], *S. Joseph Gladwin*[5], and *S. Pravin Kumar*[6]

[1] *Assistant Professor, ECE, Saveetha School of Engineering, SIMATS, Chennai*
[2] *Associate Professor, ECE, Saveetha School of Engineering, SIMATS, Chennai*
[3] *Associate Professor, ECE, SITAMS, Chithoor, AP*
[4] *Professor, ECE, Vel's University, Chennai*
[5] *Associate Professor, ECE, SSN College of Engineering, Chennai*
[6] *AI Engineer, Smartail Pvt Ltd, Chennai.*

3.1 Introduction

WANs operate with the same components as used in wired networks. But, in WANs, data are transferred over air medium. Hence, it has very poor security. In recent years, the growth of network-based service has been incredible. Therefore, network security becomes one of the important problems in cyber world. Bace and Mell [1] define *intrusion* as "efforts to compromise the confidentiality, dependability, and simplicity, or to avoid the security mechanisms of a computer or network." Using this principle, an *intrusion detection system* (IDS) can be defined as a software or hardware used to detect efforts to compromise the privacy, reliability, or accessibility of a network, or to evade the security contrivances of a network [2].

Typically, multiple nodes exist in an area, and all nodes share the same wireless medium. Therefore, there are chances for data hacking or attacks. Several security measures are available to detect and control these attacks. In general, these attacks are classified as *active* and *passive attacks*. In passive attack, the attacker passively listens to the packets in wireless medium and does not modify anything, whereas in active attacks, like wormhole attack, black hole attack, and hello flood attack, the attacker indulges in malicious action in addition to passively listening to the data. Thus, more and more attention has been focused on active attacks.

To protect our networks from unauthorized or malicious actions, we need effective IDSs. Generally, these IDSs are grouped under two categories—*signature-based detection system* and *anomaly-based detection system*. Using the previous database, a signature-based IDS employs pattern matching techniques to detect attacks, whereas an anomaly-based IDS trains itself and carefully sets the threshold value to detect attacks, thus making their deployment more complex [3]. Furthermore, new attacks are also detected by anomaly-based

Sensor Data Analysis and Management: The Role of Deep Learning, First Edition. Edited by A. Suresh, R. Udendhran, and M.S. Irfan Ahmed.
© 2021 John Wiley & Sons Ltd. Published 2021 by John Wiley & Sons Ltd.

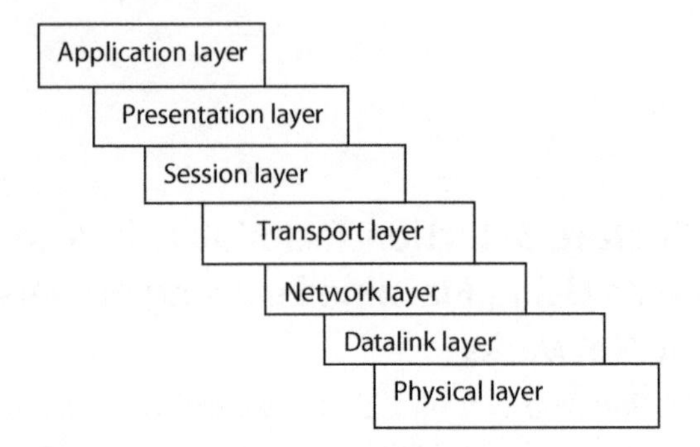

Figure 3.1 Layers of OSI reference model.

Table 3.1 Types of attacks in network layer.

Network layer	Spoofed attack Selective forwarding attack Sink hole attack	Detection of min route [6]
	Sybil attack	Identity certificates [7]
	Wormhole	DAWWSEN proactive routing protocol [8], suspicious node detection by signal strength
	Hello flood attack	Suspicious node detection by signal strength
	Acknowledgment spoofing attack	Encryption, authentication, monitoring

detection systems. Usually, the IDS uses different layer features to detect the attack. Smart selection of the number and type of features improves the accuracy of IDSs, which also thereby minimizes detection time.

A WAN constructed using 802.11 technology [4] is used in this study. The International Organization for Standardization (ISO) gives the reference model for the wireless network architecture, that is, a seven-layer open system interconnect (OSI) model. The network function of each layer is represented in Figure 3.1. This is a very handy model. However, most of the attacks happen in the lower layers.

The network layer is responsible for secure communication, routing, firewall, and address resolution protocol (ARP)/broadcasting monitoring. So, the attackers are very much interested in attacking the network layer and hacking the information easily through IP address spoofing or declaring false information to the neighbor nodes. Thus, this network layer is more vulnerable to attacks [5]; hence, more research is being conducted in the development of IDS to detect intrusions in the network layer. In network layer attacks, the wormhole attack and black hole attack are common and popular attacks. So, this work deals with the detection of wormhole attacks (Table 3.1).

3.2 Related Works in the Detection of Intrusion

To better understand existing IDSs and the techniques adopted to develop an effective system, an exhaustive literature review was done, and the same is presented here. To improve detection accuracy, from the vast varieties of features, the features having high influence on attacks are considered for the detection of intrusion. Some of the feature-minimizing techniques are discussed in the following text.

Devendra and Bedi [9] proposed a feature-based IDS system. The features are selected based on threshold values. Some of the main factors which affect IDS performance in mobile ad hoc networks (MANETs) are motion of the devices, type of construction, type of routing protocol, and detection method used, which are mainly to classify among the normal node and the attacker node. For identifying denial-of-service (DoS) assaults and reducing the false alert rate, two characterization methods—specifically, SVM and rule-based classification—have been utilized for viable grouping of the knowledge discovery in databases (KDD) informational collection.

Erasing immaterial and excess components facilitates a speedier formulating and testing process, to have less advantageous applications and also to keep up high reduction rates. The adequacy and practicality of the component choice model were confirmed by a performance analysis on KDD intrusion detection dataset. According to Sen and Dogmus [10], the networks are simulated by the network simulator 2 (NS2) tool. Here, the mobility configurations of the nodes were simulated by the random waypoint model. Different types of network settings were created with different movement levels and traffic loads. Here, 50 nodes were placed with transmission control protocol (TCP) traffic, in a topology of $1000 \times 500\ m^2$. Overall, 26 features were used in the attack detection. Here, a support vector classifier with regularization (C-SVC) algorithm was used for the detection. For normal cases 0.005 and for abnormal cases 0.1 weights were given. Here, the detection rate was more for medium rate, and for medium traffic it was 97.08%. Genetic algorithm was used in the feature selection. Features were selected based on fitness value (fitness = detection rate − false positive rate). Now, the detection rate was improved to 99.67%.

Kabiri and Aghaei [11] presented an IDS for ad hoc networks. They used PCA for the selection of features, since the inappropriate features degrade the performance of the detection system. Performance reduction occurs both in speed and predictive accuracy. Here, selection and analysis of the network features use PCA. The performance of various experiments, normal and attack states, is simulated, and the results for the selected features are analyzed. Hi et al. [12] used KDDCUP99 dataset with 41 features. They used SVM multi-class classifier to detect normal and attacked data. Here, radial basis function (RBF) kernel function is used. By using a means clustering algorithm, different types of data samples are partitioned into clusters. Nineteen features are selected by using glomerular filtration rate (GFR). The accuracy is 98.62%.

Devendra and Bedi [9] did a survey to discuss the predictable classification engines that are major applicants for the study of IDSs. They also reviewed the different intrusion detection approaches. Also, they presented a survey of the main IDS, and then categorized these mechanisms as either trust value that can deal with threshold values or as feature-based that deal with the reduction step called *preprocessing*. Their study is an evidence that the performance of the IDS is dependent on the selection of features. If the required features are not selected properly, then it leads to a high misclassification rate.

Tesfahun and Bhaskari [13] presented the synthetic minority oversampling technique (SMOTE)-based IDS and used the feature selection based on information gain (IG). The random forest (RF) algorithm is used as a classifier in the classification of normal and attacker nodes. The result shows that this classifier with SMOTE- and IG-based feature selection gives an efficient and effective IDS. For this IDS, the detection rate and false positive rate (FPR) with feature selection are 96.3% and 0%, respectively. And without feature selection, these are 96% and 0%, respectively. The precision rate is 99% with and without feature selection IDS.

Hi et al. [12] proposed a gradual feature reduction method called SVM and ant colony algorithm along with SVM to judge whether the node is normal or abnormal based on Matthews correlation coefficient. It achieves 0.861161 accuracy, which is less compared to other methods. Wenying [14] proposed support vector method and clustering-based self-organized ant colony algorithm to detect the behavior of networks. This research work shows that combining support vectors with ant colony (CSVAC) outperforms SVM in terms of both classification rate and detection parameter.

From the review, it is understood that a large number of network layer features make the IDS less accurate and slow. Hence, an IDS designed using minimum network layers is preferable. It was also found that the random forest and PCA methods were very effective when used for feature selection. Similarly, K-means clustering algorithm and one class SVM were effective in the classification of normal and abnormal nodes. As there still are many undetected intrusions, a faster and more efficient IDS is required to protect the networks from malicious attacks in this Information Age.

3.3 Framework of Wormhole Detection in WANs

The WAN is more vulnerable to network layer attacks. In these attacks, the detection of wormhole attacks is the most difficult task. The following section elucidates the framework used in the present research work to detect wormhole attacks. The simulation environment used to build the IDS is discussed in detail as well.

3.4 Impact of Wormhole Attack in WANs

The wormhole attack is an unadorned MANET routing hazard. Because it is easy to launch, it can be launched in different ways. This attack is very difficult to detect and can cause major communication interference. Wormhole nodes create a false route that is shorter than the original route within the network. This can complicate routing mechanisms which depend on information about distances between the nodes within the network [15]. This attack system has one or more attacker nodes and creates a tunnel between them. This tunnel creates an illusion that the two end points of the tunnel have a very short distance between them. The attacker node captures the packets from source and transmits them to another reserved node which distributes them loyally. This attack can be launched easily.

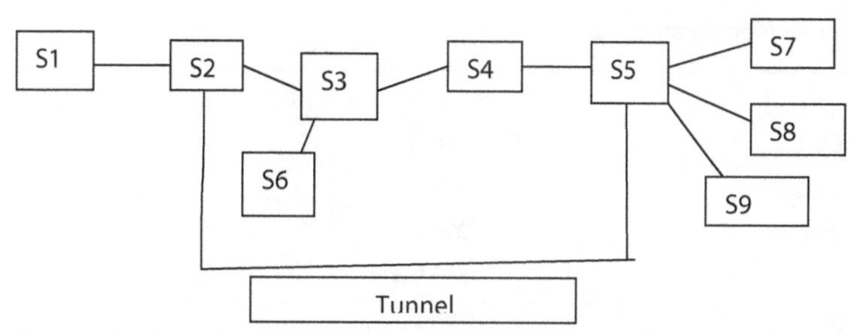

Figure 3.2 Wormhole attack.

Here, the attacker, without having knowledge of the network or nodes or cryptographic mechanisms, can hack the data easily. The tunnel may be a wireless higher frequency link or wired link. Figure 3.2 shows the wormhole attack construction.

Untruthful route generation, dropping of packets wholly or selectively, and the attraction of large amounts of data traffic are the major effects of wormhole attack.

S2: destination point, S5: origin point.

3.5 Existing IDS Techniques to Detect Wormhole Attacks

The main purpose of the existing research work is to protect the network from wormhole attacks with the advancement of ad-hoc on-demand distance vector (AODV). To compare the performance parameters, some of the works and their techniques are listed in Table 3.2.

Table 3.2 Intruder detection system for wormhole attacks.

Author	Techniques	Limitations
Chiu and Lui [16]	Delphi technique	Very high rescheduling of one hop propagation; false alarm rate not discussed
Chen et al. [17]	Localized algorithm	Detects wormhole attacks in cases where there is no packet loss
Özdemir and colleagues	Trust-based model	Works well only when both time- and trust-based modules are combined
Bensaou and colleagues	Statistical analysis of multipath (SAM)	Showed improvement in false alarm rate
Lee and colleagues	Proposed multistep multiclass IDS (MMIDS)	(a) One-class SVM to detect normal and abnormal data from the dataset (b) Multiclass SVM classifies attack data into one of DoS, remote to local (R2L), user to root (U2R), and probing attacks (c) Fuzzy adaptive resonance theory (ART) clusterizes each attack

Table 3.3 Simulation parameters.

Simulation area	(1000×1000) m^2
Node coverage	250 m^2
Number of nodes	100
Simulation time	200 s
Size of packet	512 bytes
Initial energy	100 J
Type of media access control (MAC)	802.11
Number of attacker nodes	2–10
Model of the antenna	Omnidirectional
Number of traffic	1–5
Application used	CBR

3.6 Simulation Environment Used in the Present Work

In the present work, NS2 is used for simulation. Also, 1000×1000 m^2 simulation area is used with 100 nodes, which are used in random positions. Here, AODV algorithm is used. Node 10 is selected as wormhole source node, and node 45 is used as wormhole sink node. Between these nodes, a tunnel is applied. Table 3.3 presents the parameters used in this simulation work. Here, a constant bit rate (CBR) is used for a 250 m^2 configured area, and the packets are transmitted in a randomly chosen speed from random source to random destination. The pause time (200 seconds) is used here.

In this work, 12 network layer features are used for IDS to detect the wormhole attack. These network layer features are collected from the abovementioned simulated environment.

3.7 Data Collection

This present work detects the wormhole attack. Here, the attacker node receives all packets from the sender, which are present in the network. Here, there is a remarkable variation in the percentage of route change (PRC), percentage of hop count change (PHC), maximum changes occurred in sequence number (Max_Seq), maximum changes in hop count (Max_Hop), average difference in the sequence number (Diff_Seq), and average difference in the hop count (Diff_HC). The percentage of delay change (PDC) computed from the previous history of the network is also affected. In intrusion state, the attacker node has high receiver power (percentage of receiver power change, PRPC). Here, the attacker mostly or selectively drops the packets. Therefore, percentage of drop ratio change (PDRC) is high, and percentage of packet sent count change (PPSC) is also affected. In the presence of attacks, features such as percentage of neighbor count change (PNCC) and percentage of average

Table 3.4 Network layer features.

Symbol	Name	Network features
F1	PRC	Percentage of route change
F2	PHC	Percentage of hop count change
F3	Max_Seq	Maximum changes occurred in sequence number
F4	Max_Hop	Maximum changes occurred in hop count of the WAN
F5	Diff_Seq	Average difference in the sequence number
F6	Diff_HC	Average difference in the hop count
F7	PDC	Percentage of delay change
F8	PRPC	Percentage of receiver power change
F9	PDRC	Percentage of drop ratio change
F10	PNCC	Percentage of neighbor count change
F11	PDNCC	Percentage of average difference in neighbor count with all neighbors changed
F12	PPSC	Percentage of packet sent count change

difference in neighbor count with all neighbors changed (PDNCC) also suffered with distinguished discrepancy.

Therefore, for the present study, these 12 network layer features (Table 3.4) are considered for the design of IDS in the detection of wormhole attacks.

3.8 Design of Learning Model for Feature Selection in IDS for WANs

The number of features in the network layers in the WAN should be minimum for an efficient and reliable IDS. During literature review, it was found that both RF method and PCA method were very effective in feature selection. Consequently, these two methods were used in the present study for minimizing the number of features in the network layer. The relationships between the features were analyzed, and thereby the influence of feature selection in the IDS was studied and presented here.

After designing the network with attackers, audit data were collected by analyzing the log files generated by the NS2 simulation environment mentioned in Table 3.3. As discussed earlier, the IDS for the WAN built with minimum network layer features will be more effective. Typically, 41 features are available in the network layer. However, in the present study, we have collected 12 network layer features, and the same are presented in Table 3.4.

As the number of features taken for the analyzation increases, the model interpretability decreases. Therefore, it leads to the reduction of the overall training speed and performance of the IDS. Thus, it is necessary to select fewer features from the data that are most important for the analyzation of the problem. Thus, feature selection becomes very important. The process of selecting a subset of relevant features for use in model construction is called *feature selection*. Nowadays, machine learning models are used widely in feature selection in WANs. In this work, two machine learning models are used for feature selection, namely, random forest method and PCA method.

3.8.1 Random Forest Feature Selection Technique

Tin Kam Ho initially created the random forest method. Later, Leo Breiman and Adele Cutler developed its extension. This algorithm combines the Breiman's bagging idea and random selection of features. It constructs a group of decision trees with controlled variance. Random Forest technique is a flexible method that easily implements machine learning algorithms even without hyperparameter tuning. Because of its simplicity and the fact that it can be used for both classification and regression tasks, random forest algorithm can be used for both supervised and unsupervised learning models. This algorithm creates a forest with several trees. In general, more trees make the forest more robust. The higher number of trees makes the outcomes more accurate.

Generally, the random forest method is classified under two categories, namely, supervised learning method and unsupervised learning method.

The random forest method is the best method for feature selection in IDSs. It identifies the best significant features out of the available features from the training dataset. The overfitting problem will never occur. The random forest algorithm can be used for classification and regression tasks.

3.8.2 PCA Feature Selection Technique

The PCA technique decreases the dimension or number of features of a dataset, which consist of many variables that are correlated to each other. This is done by converting the scaled dataset to a new set of variables. The first principal component maintains the maximum variation present in the original components. These are eigenvectors of a covariance matrix, and, hence, they are orthogonal. When we move from first principal component to last principal component, the variation present in the principal components reduces. Some of the commonly used terms in PCA are as follows:

Correlation: This shows the relationship between two variables. The values range from −1 to +1. The +ve values indicate that, when one variable increases, the other variable also increases. The −ve value indicates that, when one variable decreases, the other variable also decreases.

Modulus value: This provides information about the strength of relationships between the features.

Orthogonal: Uncorrelated with each other. The correlation factor is 0.

Eigenvectors: Consider a non-zero vector v that is an eigenvector of a square vector $\bar{A}$. $\bar{A}$ v is a scalar multiple of v, then

$\bar{A}\,v = \lambda\,v$

Here, the λ is an eigenvalue associated with v.

Covariance matrix: This matrix consists of the covariance between the pair of variables. In the matrix, $(i, j)^{th}$ variable is the covariance between i^{th} and j^{th} variables.

3.8.3 Procedure to Implement PCA in IDSs

Step 1: The dataset to be normalized. Here, the dataset mean is zero.
Step 2: Construction of covariance matrix:
$\mathrm{Var}[Z_1] = \mathrm{Cov}[Z_1, Z_1]$
$\mathrm{Var}[Z_2] = \mathrm{Cov}[Z_2, Z_2]$

$$\text{Covariance matrix} = \begin{pmatrix} \mathrm{Var}[Z_1] & \mathrm{Cov}[Z_1, Z_2] \\ \mathrm{Cov}[Z_2, Z_1] & \mathrm{Var}[Z_2] \end{pmatrix}.$$

Step 3: Calculate the eigenvalues and eigenvectors of the covariance matrix. The eigenvalue λ of a corresponding eigenvector v is found by solving $(\lambda I - A)\,v = 0$. Here, λ is the eigenvalue of matrix A. I is the identity matrix.
Step 4: To select the eigenvectors, for *n* variable dataset, we have *n* number of eigenvectors and eigenvalues. To reduce the number of features, the variables are selected in the order of highest eigenvalues:

$$\text{Feature vector} = \{\mathrm{eig}_1, \mathrm{eig}_2, \mathrm{eig}_3, \&\}$$

Step 5:Construction of principal component (PC) matrix:

$$\text{New dataset} = \text{feature vector}^T \times \text{scaled data}^T$$

T indicates the transposed vector; new dataset is a matrix with principal components; and feature vector set is a matrix with eigenvalues. Scaled dataset is a scaled version of the original dataset.

3.8.4 Data Collection

The data for the 12 network layer features are generated using the simulation environment mentioned in Table 3.3. The 12 network layer features used in the present study are given in the following text:

1) Route changed in percentage.
2) Hop count changed in percentage.
3) Maximum changes occurred in sequence number.
4) Maximum changes occurred in hop count of the WAN.
5) Average difference in sequence number.
6) Average difference in hop count.
7) Delay changed in percentage.
8) Receiver power changed in percentage.
9) Drop ratio changed in percentage.
10) Neighbor count changed in percentage.

11) Average difference in neighbor count with all neighbors changed in percentage.
12) Packet sent count changed in percentage.

The huge amount of data involved diminishes accuracy and increases processing time. Therefore, with the help of the feature selection method, four features with high impact were extracted and used to build the IDS.

3.8.5 Existing Feature Selection Techniques in WANs

In WANs, different types of feature selection, feature extraction, and feature reduction techniques are used for intrusion detection.

3.8.6 Influence of Features in WANs

The choice of features plays an important role in the IDS of WANs. Hence, a clear understanding of the relationship between the features is very much essential for designing an effective IDS. Therefore, an attempt is made to recognize the relationship between the features using the correlation factor value.

The correlation matrix plot of the relationship between the 12 network layer features is shown in Figure 3.3. The value in each cell represents the correlation factor between the two features. The correlation factor between the features PRC and Max_seq is −0.7, whereas the correlation factor between the features PHC and Diff_Seq is −0.9. Similarly, the correlation factor between all the 12 features is computed and illustrated in Figure 3.3. High values of correlation factor of any feature imply that the influence of that feature on the WAN is relatively more. Hence, based on this correlation factor, more impactful features can be identified, and the same can be used to build an effective IDS.

With the help of a swarm plot, the status of variance of nodes, before and after feature selection, is performed as per swarm plot technique, as depicted in Figures 3.4 and 3.5. The plots were made with the features in abscissa and the threshold of features in ordinates. The data standardized using its mean and standard deviations were used for the plot. It was found that the data were normally distributed with zero mean. The features that fall out of the boundary are considered as less impactful. If all the features are considered for intrusion detection, it makes the system increasingly complex. Therefore, only features with notable variance are considered for the design of IDSs. This technique helps avoid the problem of overfitting.

The swarm plots of the four selected features—PHC, PRC, receiver power (Rxpwr), and drop ratio—are portrayed in Figure 3.5. By dropping the features having less variation score, the complexity of the problem has been considerably reduced.

3.8.7 Feature Selection by RF Technique

A smaller number of features is required because it diminishes the difficulty of the model, so a simple model is followed, which is easy to understand. The feature selection process discovers the important features from the network layer features using the training dataset. In the RF method, the data are preprocessed using standard-scaler function, and the "gini"

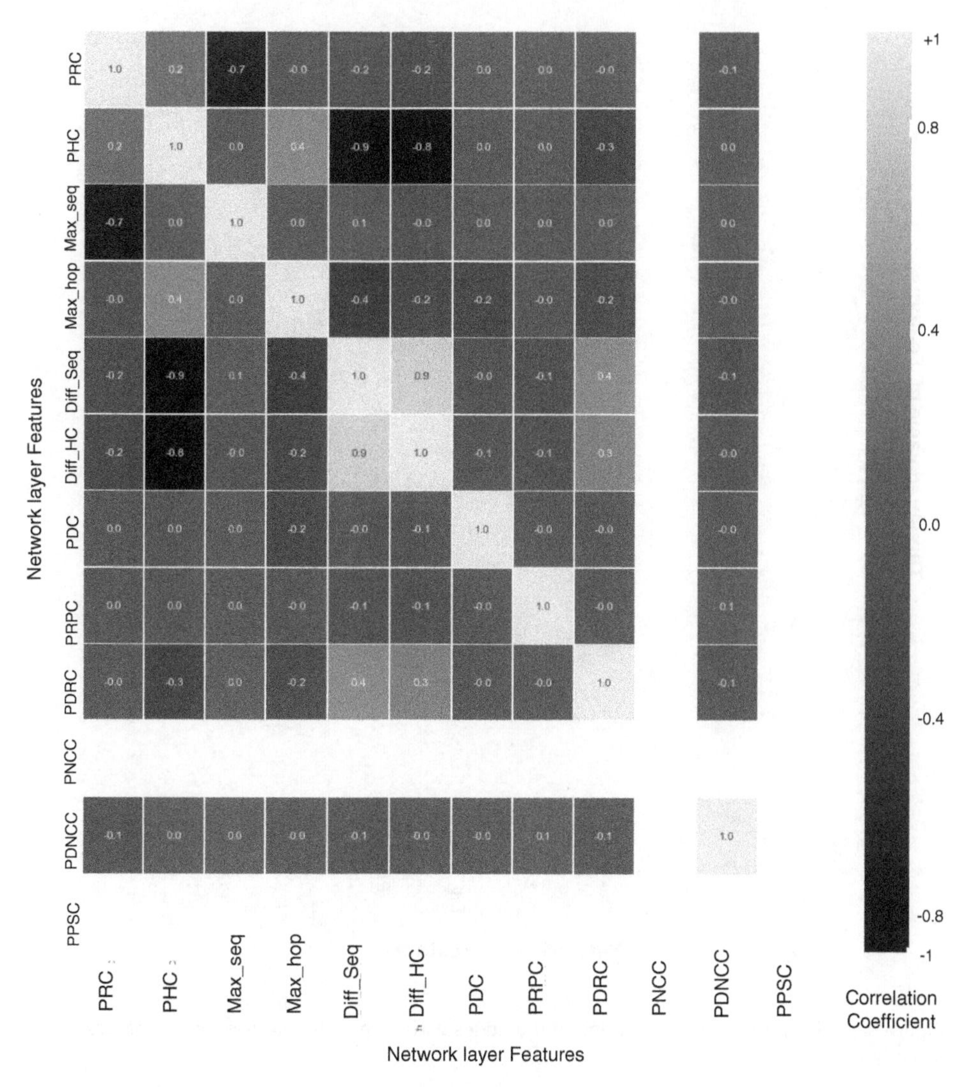

Figure 3.3 Relation between the 12 network layer features.

technique was used for data fitting. The dataset was divided into training dataset and test dataset. Here, the dataset is divided in the ratio of 60%:40% for test dataset:training dataset. The weightage of each feature is calculated using accuracy score module. The top scored features are taken for attack detection, since the impact of the feature on the ad hoc network is measured based on its score. Figure 3.6 shows the score obtained by the abovementioned 12 network layer features. From the scoreboard, the following four features were selected for the IDS:

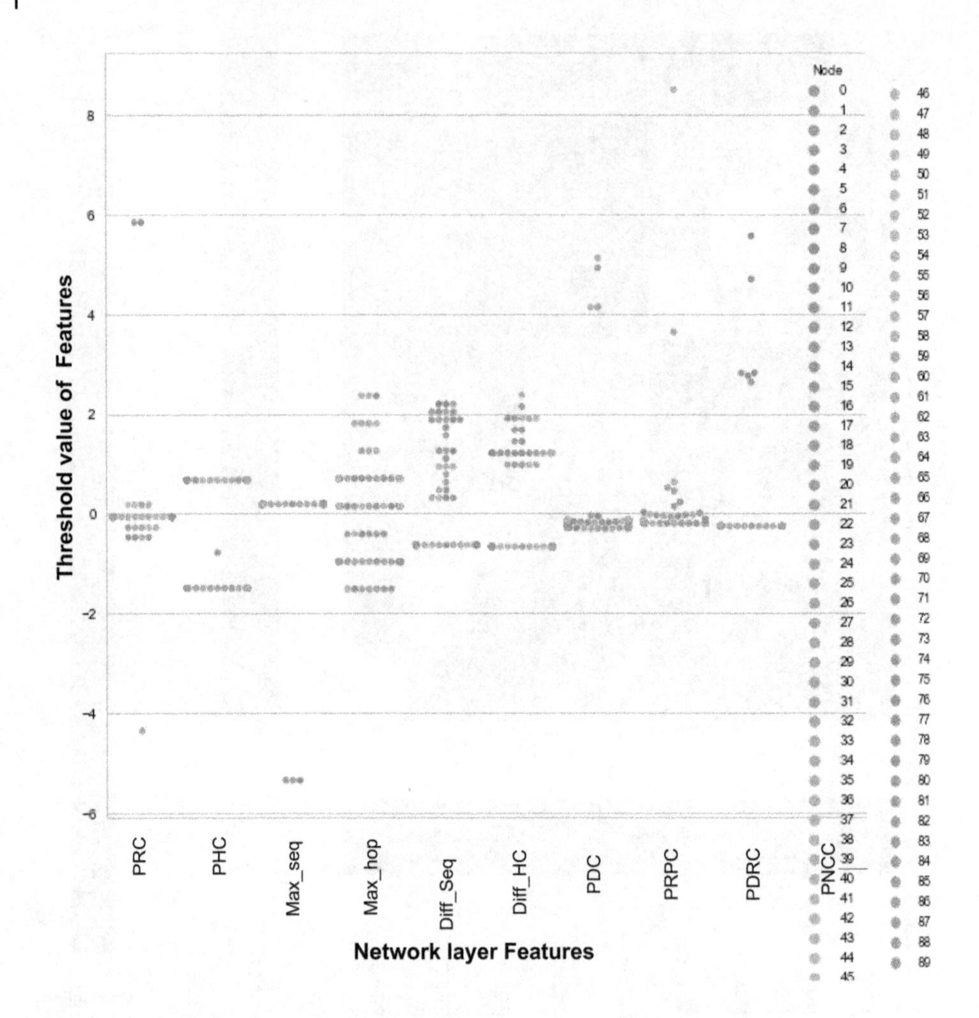

Figure 3.4 Swarm plot representation of the nodes with respect to features before feature selection.

1) Route count changed in percentage.
2) Hop count changed in percentage.
3) Receiver node power changed in percentage.
4) Drop ratio.

In this work, with the help of the abovementioned selected features, the attacks are detected using one-class SVM and K-means-cluster-algorithm-based IDS.

3.8.8 Feature Selection by PCA Technique

PCA is used to visualize data and speed up the detection process. Also, the number of inputs can be minimized using this technique. From the 12 features in the network layer,

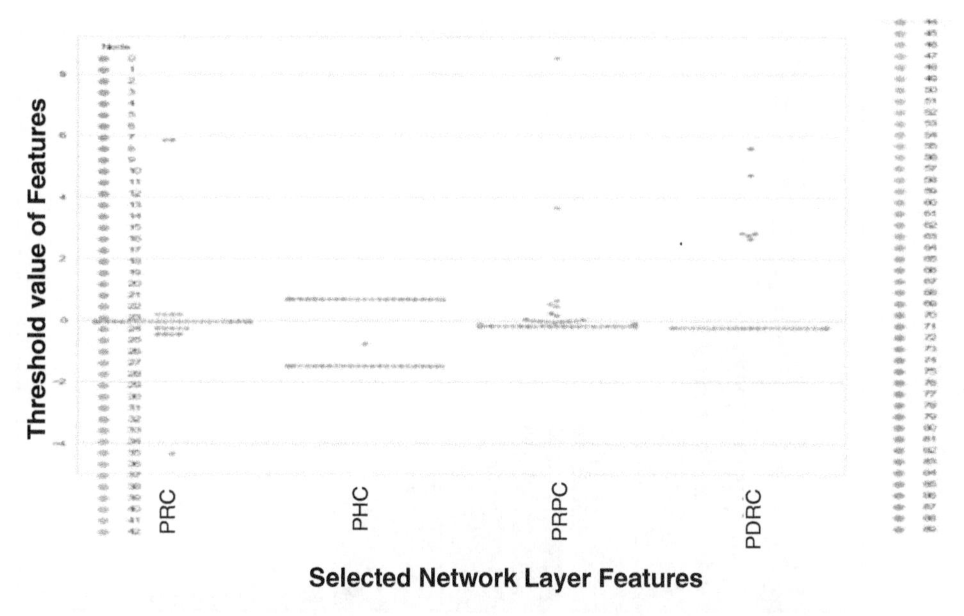

Figure 3.5 Swarm plot representation of the nodes with respect to features after feature selection.

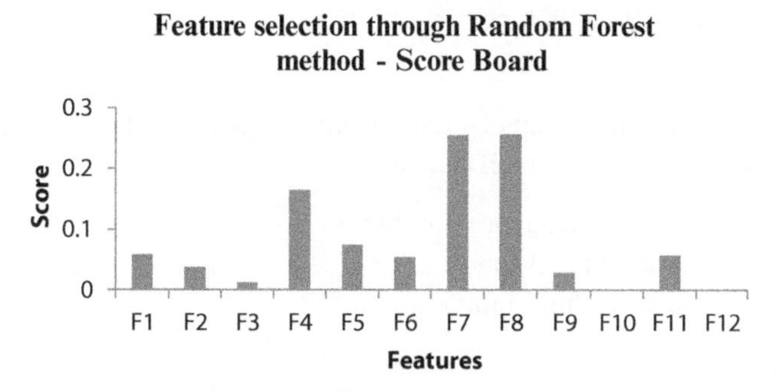

Figure 3.6 Feature selections by random forest method.

highly impactful features were selected using the PCA method. PCA detects the correlation between variables. Only if a robust correlation exists between variables does the effort to reduce the dimensionality makes sense. It retains the information, finds its directions of maximum variance in high-dimensional data, and projects it into a smaller dimensional subspace. Initially, the data are preprocessed using standard-scaler function. The covariance matrix is used to obtain the eigenvectors and eigenvalues. These eigenvalues were sorted in descending order, and the k eigenvectors corresponding to the k largest eigenvalues are identified, where k is the number of dimensions of the new feature subspace. The selected k eigenvectors are used to construct the projection matrix W. Then, the original dataset X is transformed via W to obtain a k-dimensional feature subspace Y. Finally, using the matrix Y, the variance of the principal component is computed. The

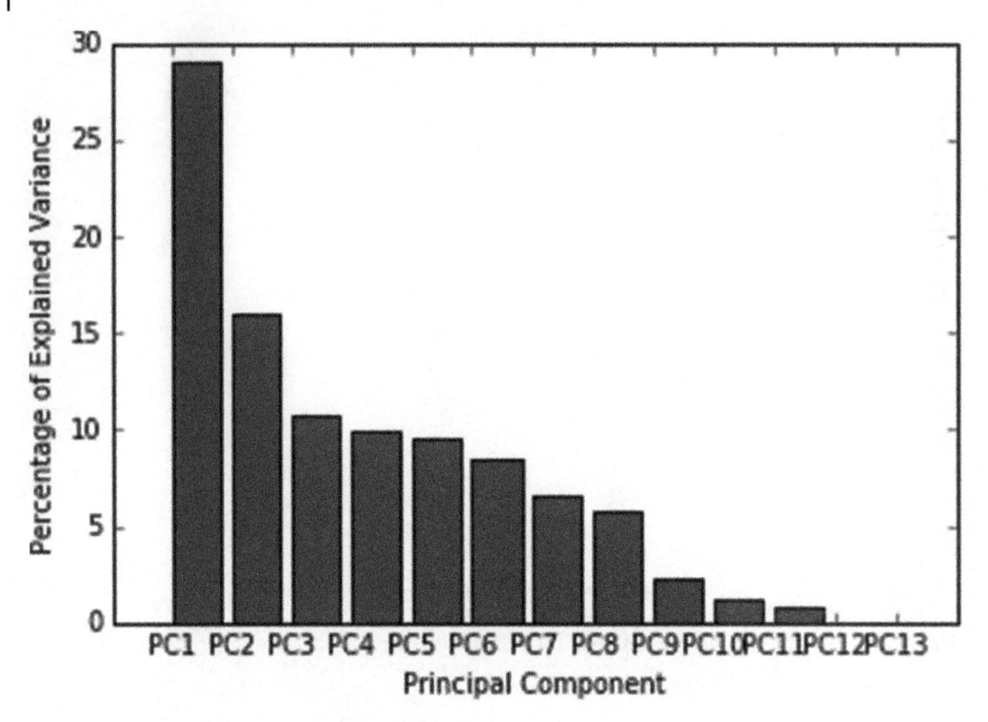

Figure 3.7 Feature selection by PCA method.

variance of each feature represents its impact on the network. The four features that are more impactful are selected for the IDS, namely:

1) Average difference in the sequence number.
2) Maximum changes in the hop count.
3) Average difference in the hop count.
4) Drop ratio.

Figure 3.7 shows the variance of all network layer features in terms of transformed components.

3.9 Design of Learning Model for IDS in WANs

An efficient machine learning model can be designed for the IDS of a WAN using various techniques. Previous research has shown that one-class SVM model and K-means cluster classifier model are more accurate in anomaly detection. This chapter demonstrates the model developed for intrusion detection using one-class SVM technique. Separate models were developed by considering scenarios both with and without feature selection. Furthermore, K-means-cluster-classifier-based machine learning models were also developed. As with the previous method, using K-means cluster classifier method, separate models were developed by considering scenarios both with and without feature selection for the detection of wormholes.

3.9.1 Machine Learning Models

The capability of processing large volumes of data within short durations makes machine learning models very popular in the Information Age. In general, machine learning languages like R, Python, Octave, and MATLAB are used to build these machine learning models. However, Python is the most practicable language because of its simplicity and effectiveness. Also, Python supports deep neural network (DNN) frameworks such as TensorFlow and café. In the present research work, both one-class SVM classifier and K-means cluster classifier are used to classify the normal and abnormal nodes in intrusion detection.

3.9.2 Methodology

In the present work, four IDSs are built with feature selection technique, and the performance metrics are compared with IDSs built without using feature selection technique. The obtained results are also compared with the performance metrics of existing IDSs. Initially, all the 12 network layer features data, without incorporating "feature selection," were used to classify normal and abnormal nodes using one-class SVM. Then, by using random forest feature selection technique, four features are selected, and the intrusion is identified using one-class SVM. Similarly, the PCA method is also used for "feature selection" and thus four features are selected, and the intrusion is identified by means of one-class SVM.

As exemplified previously for one-class SVM method, initially, all the 12 network layer features data, without incorporating "feature selection," were used to detect the wormhole using K-means cluster algorithm. Then, by using random forest feature selection technique, four features are selected, and the wormhole attack is detected using K-means cluster algorithm. Similarly, the PCA method is also used for "feature selection"; thus, four features are selected, and the wormhole attack is identified by means of K-means cluster algorithm. The flow chart of the methodology adopted in the present study for the design of IDSs is given in Figure 3.8.

3.9.3 IDS Design Using One-Class SVM Classifier

The abnormal node detection in ad hoc network is attempted using one-class SVM. This technique is used for density estimation and novelty detection. The anomalous or outliers are generally classified as point anomalies, contextual anomalies, and collective anomalies.

Point anomalies are used if an individual data instance is considered as anomalous with respect to the rest of the data (e.g., purchase with large transaction value). Contextual anomalies were adopted if the anomaly occurs at a certain time or certain region (e.g., large spike in the middle of the night). Collective anomalies were chosen if a collection of related data instances were anomalous with respect to the entire dataset, but not individual values.

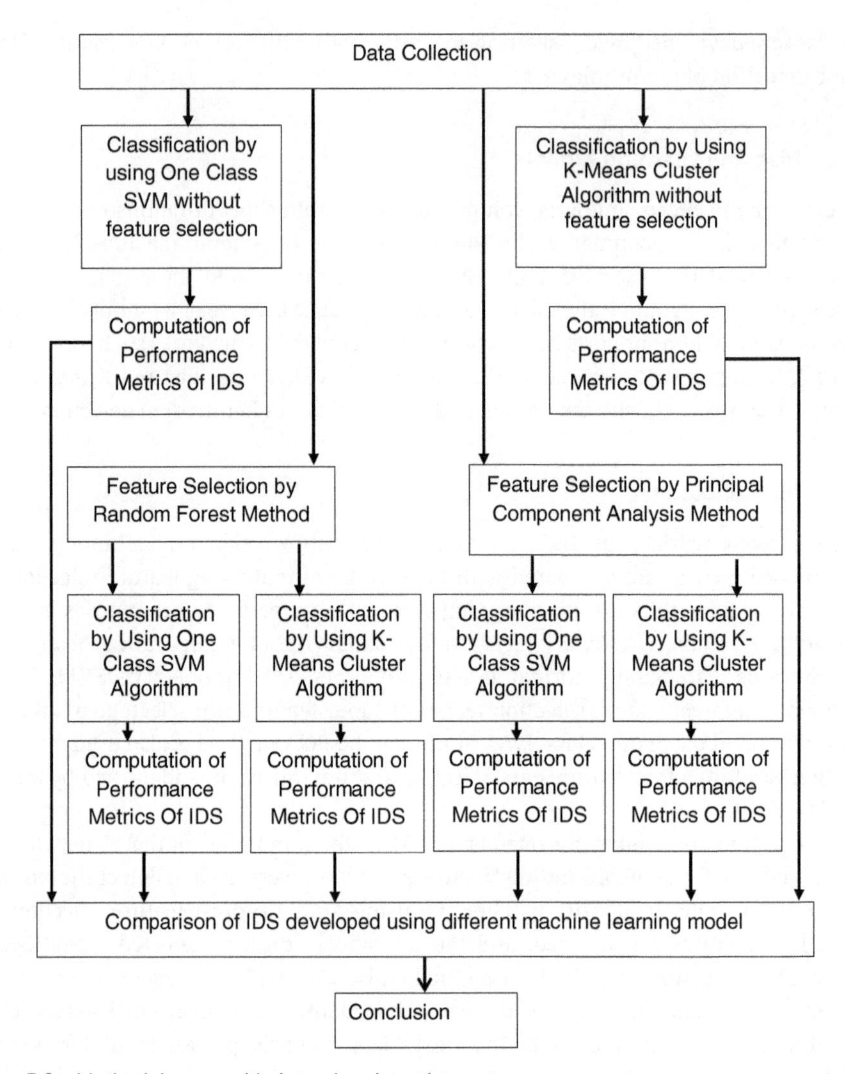

Figure 3.8 Methodology used in intrusion detection systems.

The first step is to preprocess the data. Then, through repeated iterations, the outlier fraction is selected. To train the model, 0.95 is used as the "nu function." Subsequently, the model is trained with 40% of the dataset, since the reference data for attack detection were not available. Then, the model was tested using the remaining 60% of the available data. The feature data are classified with the labels +1 or −1 to indicate whether the data are an "inlier" (normal nodes) or "outlier" (abnormal nodes). This procedure transforms our data from multiclass (multiple different labels) to one-class (Boolean label). Thus, the anomaly is detected, and the corresponding charts (Figure 3.9–Figure 3.20) are plotted. However, this procedure just identifies the intrusion, but not the type of intrusion. To identify the type of attacks, further classification of abnormal nodes is required.

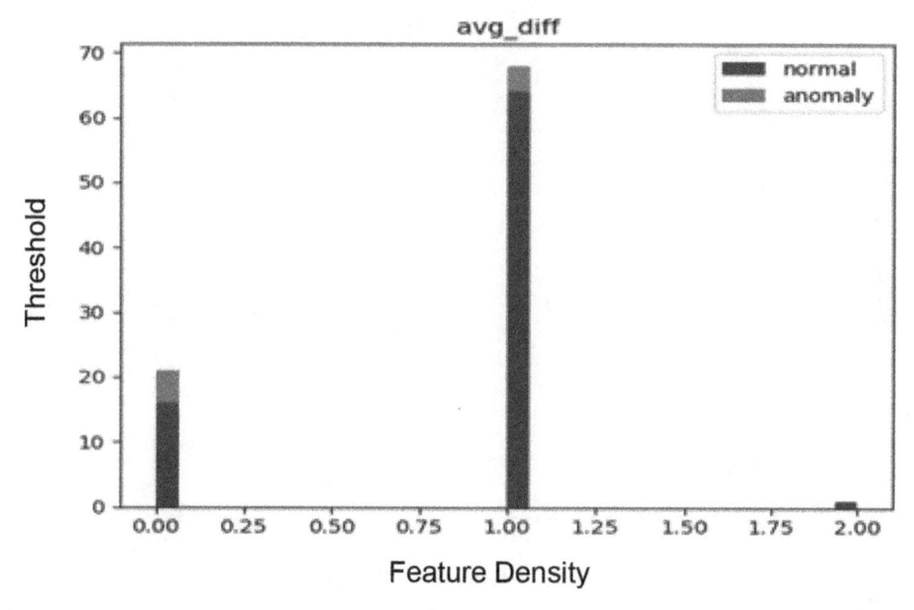

Figure 3.9 Anomaly detection with average difference in neighbor count in percentage with all neighbors changed feature in one-class SVM.

3.9.4 IDS Using One-Class SVM Without Feature Selection

This ad hoc network dataset does not have any static pattern and is an unbalanced and autocorrelated system. Figure 3.9 provides the plot between the nodes versus avg_diff (average difference in neighbor count in percentage) with all neighbors changed feature strength. The red region area gives the number of nodes that behave abnormally. The blue represents the normal behavior of nodes. From the illustration, we note that around four nodes are behaving abnormally. The abnormal node has more change in neighbor count. Therefore, the percentage of average difference in neighbor count is one of the important features in WANs.

Figure 3.10 sketches the relationship between the percentage of delay change with respect to the nodes. The abnormal nodes have more changes in delay. The plot for the number of nodes versus percentage of packet sent count changed is shown in Figure 3.11. In black hole attacks, the attacker declares itself to be the corresponding receiver, and, thereby, it receives all packets and drops. So, the original receiver sends the message repeat request to the sender, and the sender sends the messages again. Therefore, the percentage of packet sent count changed varies abnormally.

Figure 3.12 gives the one-class SVM plot for the nodes and percentage of packet drop ratio. Here, both wormhole attacker node and black hole attacker node declare themselves as receivers and receive all packets and drops. So, the attacker node has a higher drop ratio. Using this feature, the attacks on nodes can be identified easily. The plot for average difference in hop count feature with nodes is given in Figure 3.13. After intrusion, the hopping pattern behavior gets abnormal in the network. Figure 3.14 shows the detection of

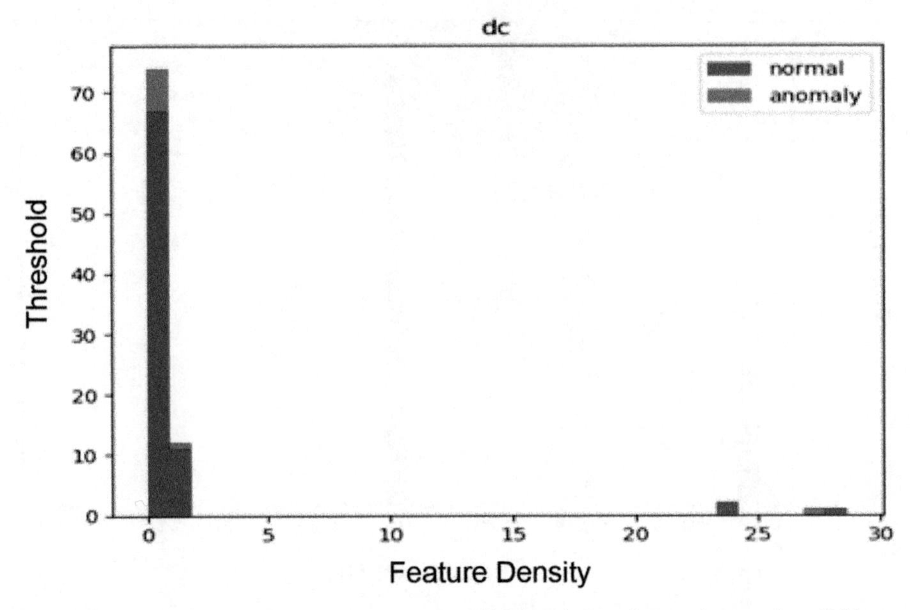

Figure 3.10 Anomaly detection using percentage of delay change feature in one-class SVM.

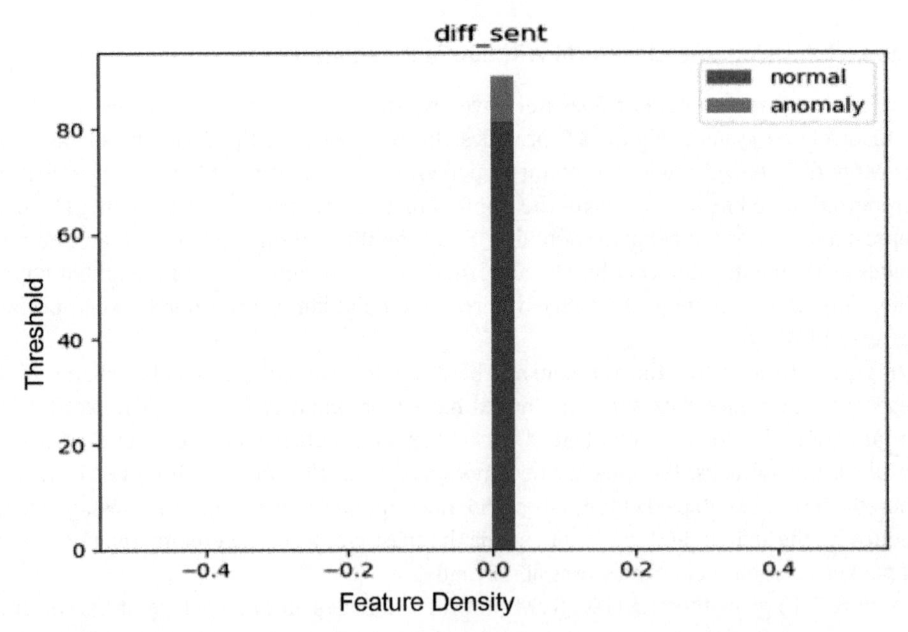

Figure 3.11 Anomaly detection using percentage of packet sent count changed feature in one-class SVM.

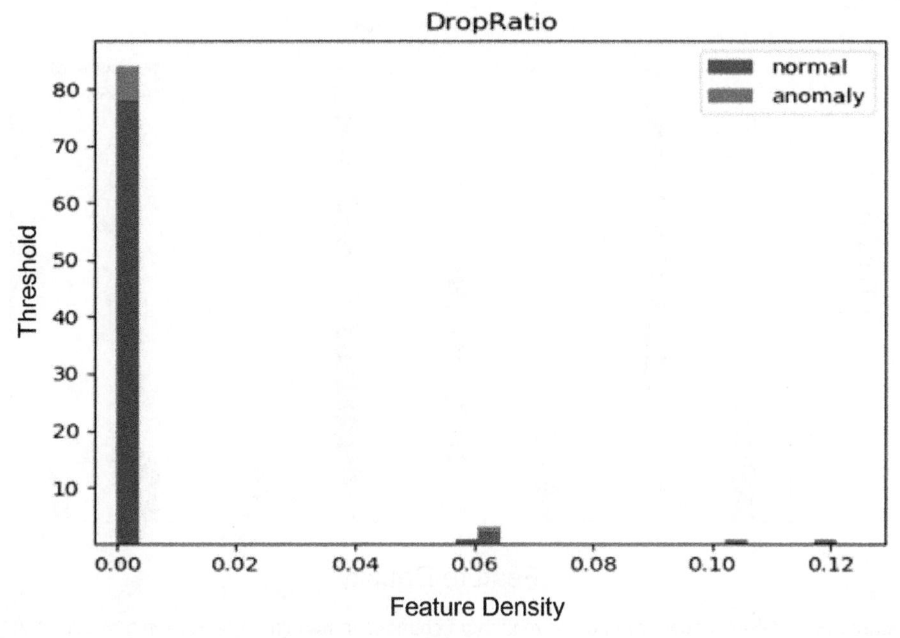

Figure 3.12 Anomaly detection using percentage of drop ratio changed feature in one-class SVM.

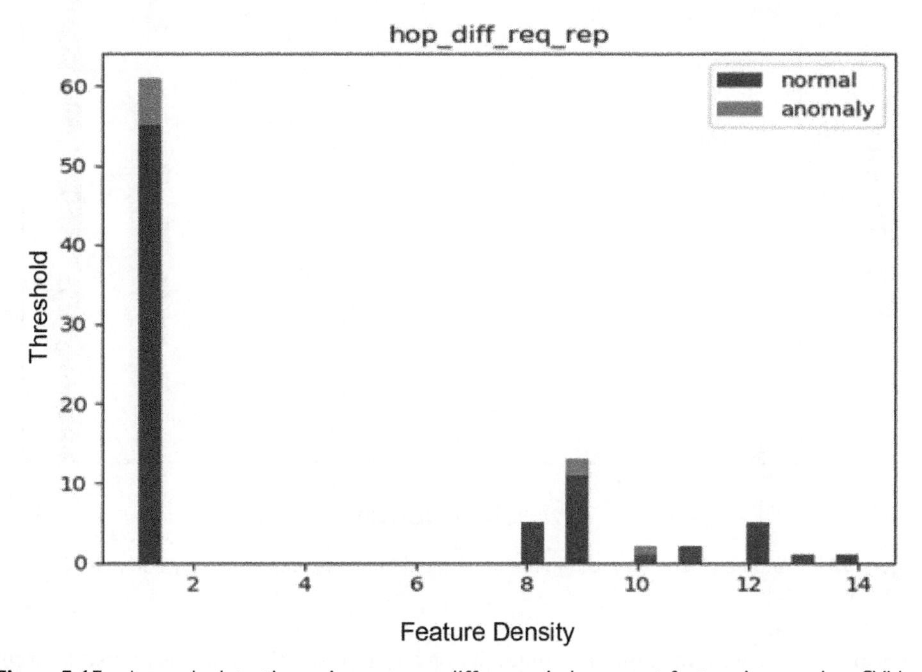

Figure 3.13 Anomaly detection using average difference in hop count feature in one-class SVM.

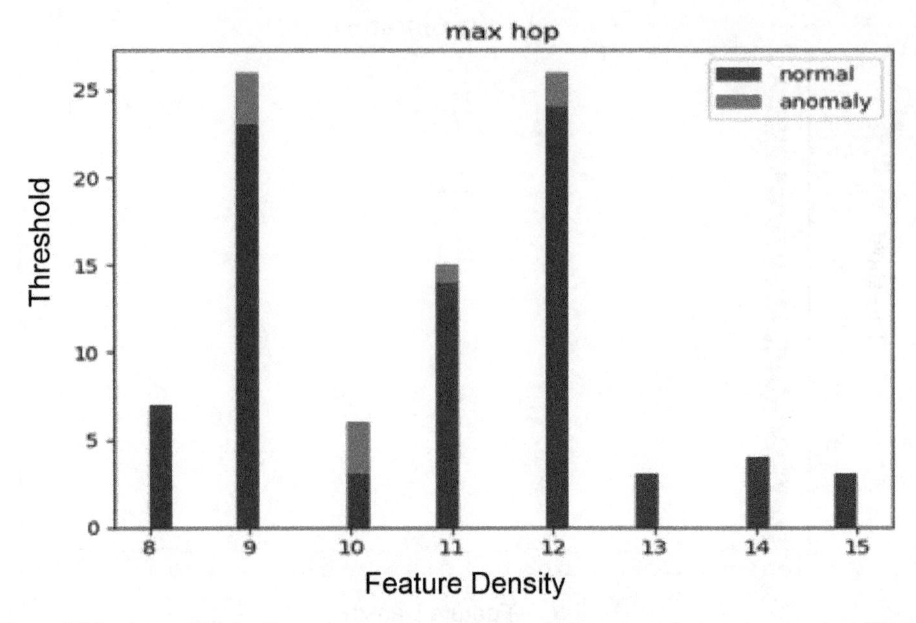

Figure 3.14 Anomaly detection using maximum changes in hop count feature in one-class SVM.

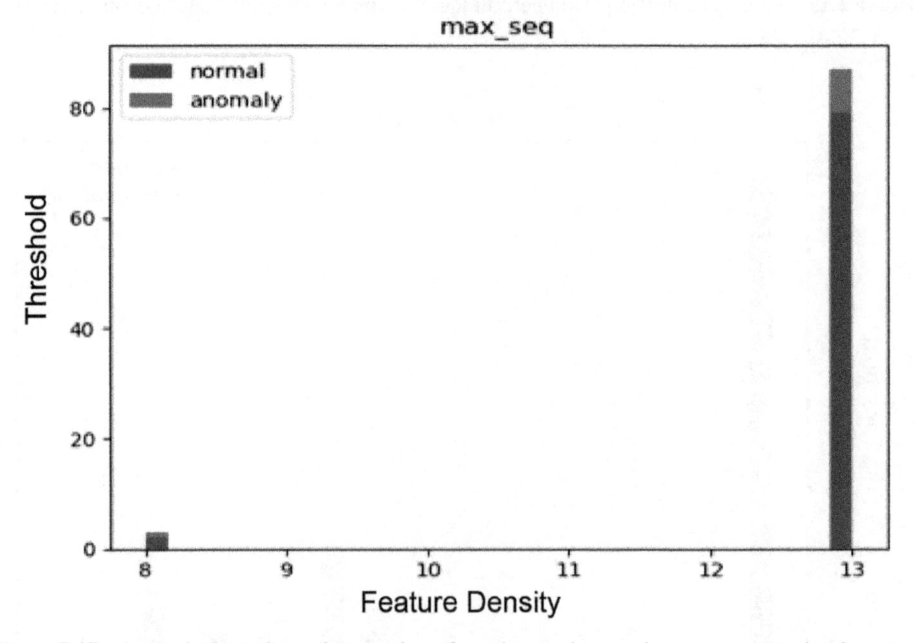

Figure 3.15 Anomaly detection using number of maximum changes in sequence number in one-class SVM.

abnormal nodes with maximum changes in hop count. Here, the sketch is plotted between the number of nodes and maximum changes in hop count. The relationship between the nodes and maximum changes in sequence number is illustrated in Figure 3.15. In the presence of the attacker, the hop count and sequence number change more randomly. The variation will be more with abnormal pattern.

The relationship of percentage neighbor counts changed with respect to nodes in one-class SVM is depicted in Figure 3.16. The number of neighbor count varies abnormally when the node acts as an attacker. Figure 3.17 shows the percentage hop count changed with respect to nodes in one-class SVM. When any node starts behaving abnormally, then the number of hop count and the sequence number of neighbor nodes change. Thus, this is also one of the important features used in intrusion detection.

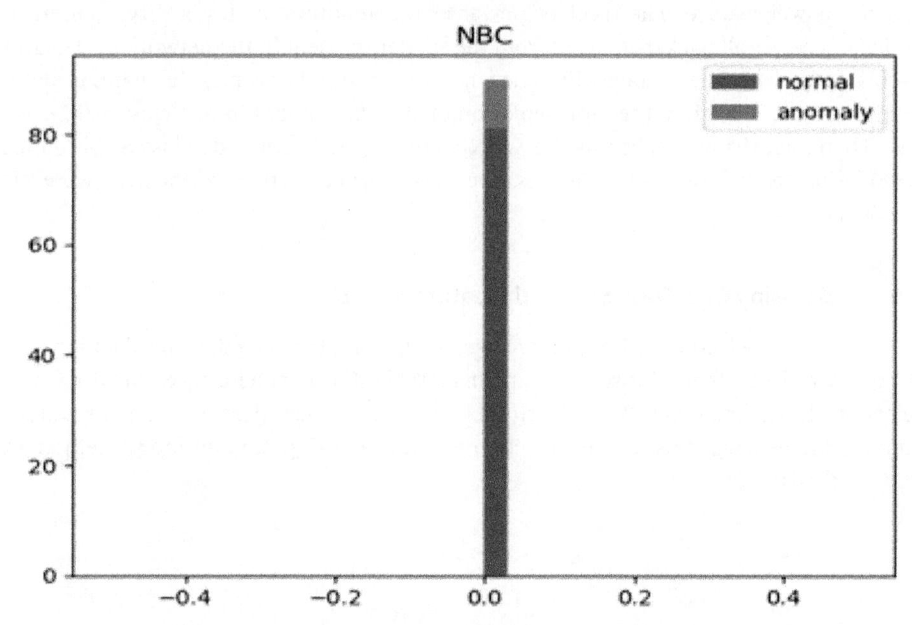

Figure 3.16 Anomaly detection using percentage neighbor count changed in one-class SVM.

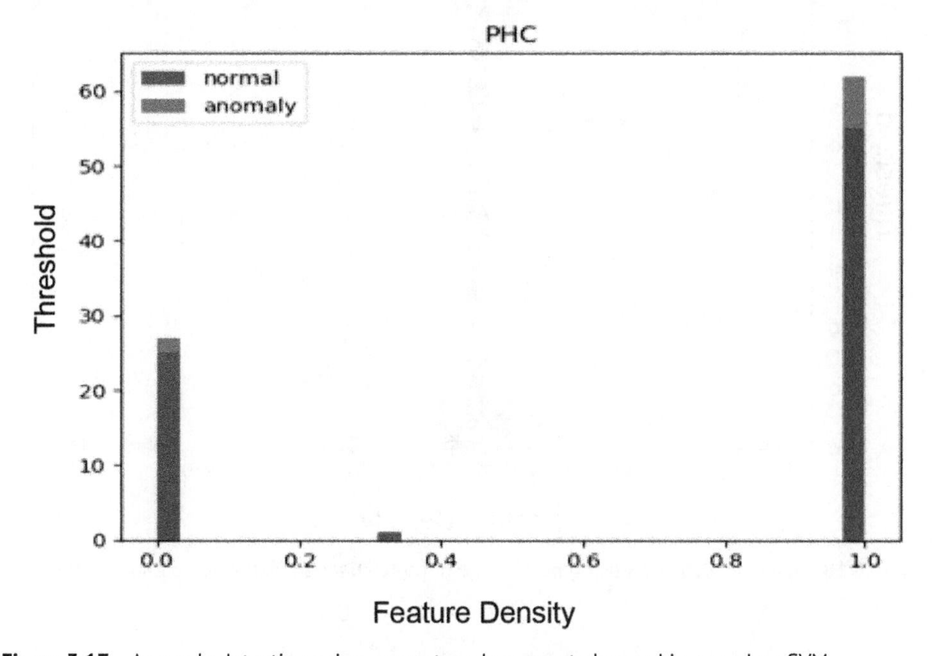

Figure 3.17 Anomaly detection using percentage hop count changed in one-class SVM.

Figure 3.18 shows the one-class SVM plot for anomaly detection using percentage route changed (PCR) feature. The graph is plotted between the number of nodes and PCR. For abnormal nodes, the PCR is very low, since the attacker routes all packets to itself by declaring itself as a receiver.

Similarly, Figure 3.19 illustrates the relation between the nodes and the percentage of receiver power changed. This represents the anomaly behavior of nodes with respect to receiver power feature. The receiver power of the abnormal nodes is very high, as the attacker receives all packets to itself from the senders present in the network by declaring itself as a receiver. The average difference in sequence number versus the nodes is plotted in Figure 3.20. This gives the representation of abnormal nodes in one-class SVM classifier. Therefore, the anomalies in the WAN were detected using one-class SVM without considering feature selection. Besides, the performance metrics of the IDS were also computed.

3.9.5 IDS Using One-Class SVM with Feature Selection

In IDSs, feature selection techniques are used to increase the detection speed and to avoid the overfitting problem. This section explains IDSs built using one-class SVM with feature selection. Here, one-class SVM algorithm is used as a classifier to detect normal and abnormal nodes, whereas feature selection is achieved using random forest method and PCA methods.

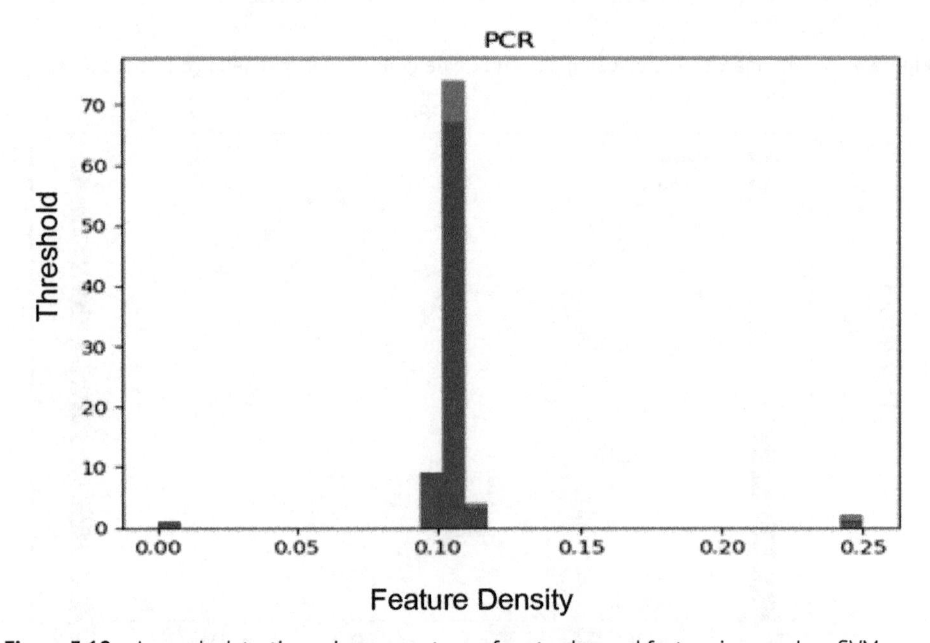

Figure 3.18 Anomaly detection using percentage of route changed feature in one-class SVM.

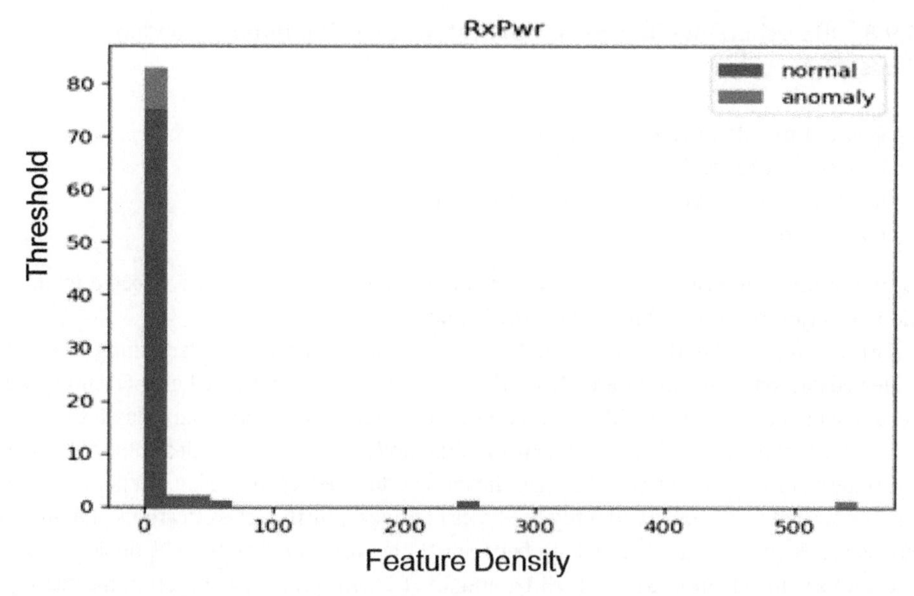

Figure 3.19 Anomaly detection using percentage receiver power changed feature in one-class SVM.

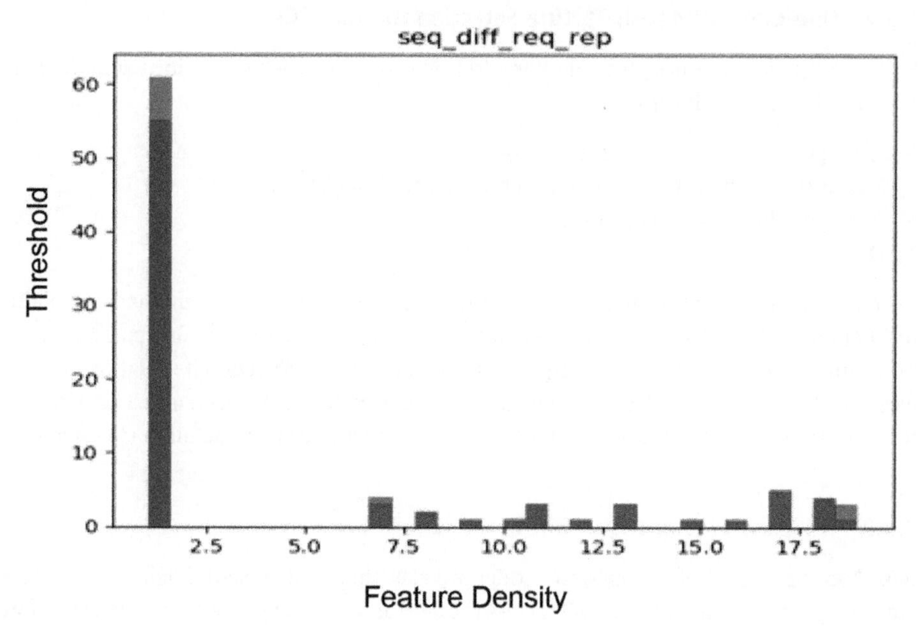

Figure 3.20 Anomaly detection using average difference in sequence number feature in one-class SVM.

3.9.6 IDS Using One-Class SVM with Feature Selection through Random Forest Method

1) Route count changed in percentage.
2) Hop count changed in percentage.
3) Receiver node power changed in percentage.
4) Drop ratio.

By using random forest method, the abovementioned four features are selected for intrusion detection from the 12 network layer features.

This IDS uses 40% of the collected dataset as training dataset and the remaining 60% of the collected dataset as testing dataset. Now, the model trains itself with the help of training dataset and learns the behavior of the nodes present in the network. The system classifies normal and abnormal nodes based on the behavior of the nodes and the properties of the features on the nodes. During the first step, the classification system detects intrusion but does not identify the type of intrusion. To identify the types of attacks, further classifications of abnormal nodes are required. The relationship between the features and nodes was depicted in the previous section (Figures 3.12, 3.17, 3.18, and 3.19). However, the performance metrics of all the developed IDSs are discussed in detail in the following chapter.

3.9.7 One-Class SVM with Feature Selection through PCA

By using the PCA method, the following four features are selected for intrusion detection from the 12 network layer features:

1) Average difference in the sequence number.
2) Maximum changes that occurred in hop count of the WAN.
3) Average difference in hop count.
4) Drop ratio.

The complete procedure adopted for feature selection using the PCA method is explained in Chapter 4. The one-class SVM classifier model is designed with the same parameters and procedure followed during the design of RF + one-class SVM IDS. The relationships between the features and nodes were portrayed in Figures 3.12, 3.13, 3.17, and 3.20. However, the performance metrics of all the developed IDSs are discussed in detail in Chapter 4.

3.9.8 K-Means-Cluster-Algorithm-Based IDS

Based on the impact of features, the nodes are classified and clustered using the K-means cluster algorithm. The number of clusters is determined through the "elbow point" shown in Figure 3.21. As illustrated in Figure 3.21, the transition drops at three; hence, the number of clusters is restricted to three.

3.9.9 IDS Using K-Means Cluster Algorithm Without Feature Selection

K-means cluster algorithm notices the groups in the data. Based on the similarity of the features, this algorithm works iteratively to assign each data point to one of the K groups.

Here, the centroids of the K cluster are used to label new data. These centroids are the collection of feature values. These feature values qualitatively take the kind of group each cluster is associated with.

For this algorithm, the inputs are the number of clusters "K" and the dataset. The algorithm starts with initial estimates, for the K centroids. This may be randomly generated or randomly selected from the dataset. This algorithm iterates between the following two steps.

Data assignment step: Each centroid defines one of the clusters. Each data point is assigned to its nearest centroid, based on the squared Euclidean distance. The data point x is assigned to a cluster based on Equation (3.1), where C is the dataset and c is the collection of centroids:

$$\underset{c_j \in C}{\arg\min}\; dist\left(c_j, x\right)^2 \tag{3.1}$$

Centroid update step: Let S_i be the centroid of the set of data point assignments for each ith cluster. Now the centroids are recomputed by taking mean of all data points assigned to that centroid's cluster using Equation (3.2):

$$C_i = \frac{1}{\left|S_i\right|}\sum\nolimits_{x_i \in S_i} x_i \tag{3.2}$$

This algorithm finds clusters and dataset labels for a prechosen K. The number of clusters K is obtained using the "elbow point" technique (Figure 3.21). The mean distance between data points and their cluster centroids is used as a metric to compare the results. Here, the complete data fall along the continuous feature ranges within one single group. From the effects of features falling on the clusters, we can detect the type of intrusion. Therefore, this technique can be used to detect the active attacks, even though they are not having any previous history, and thereby they can be used to find the new attacks.

Here, the 12 network layer features are taken for the identification of intrusion detection. The classification of clusters using K-means cluster algorithm without feature selection is

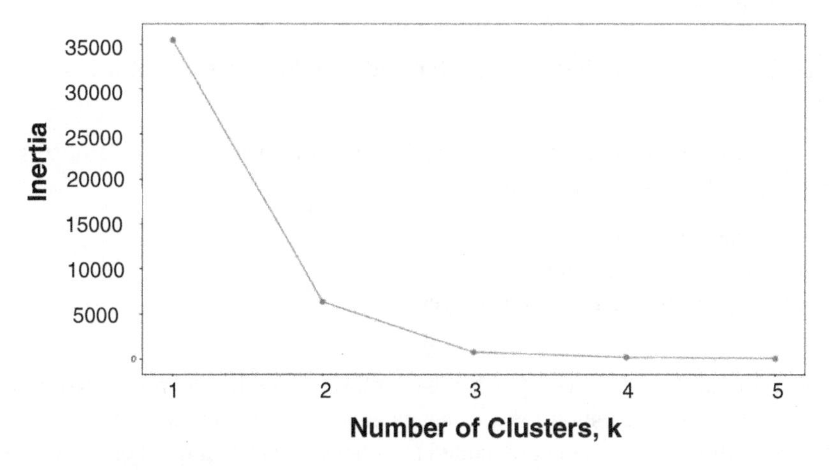

Figure 3.21 Elbow point determination.

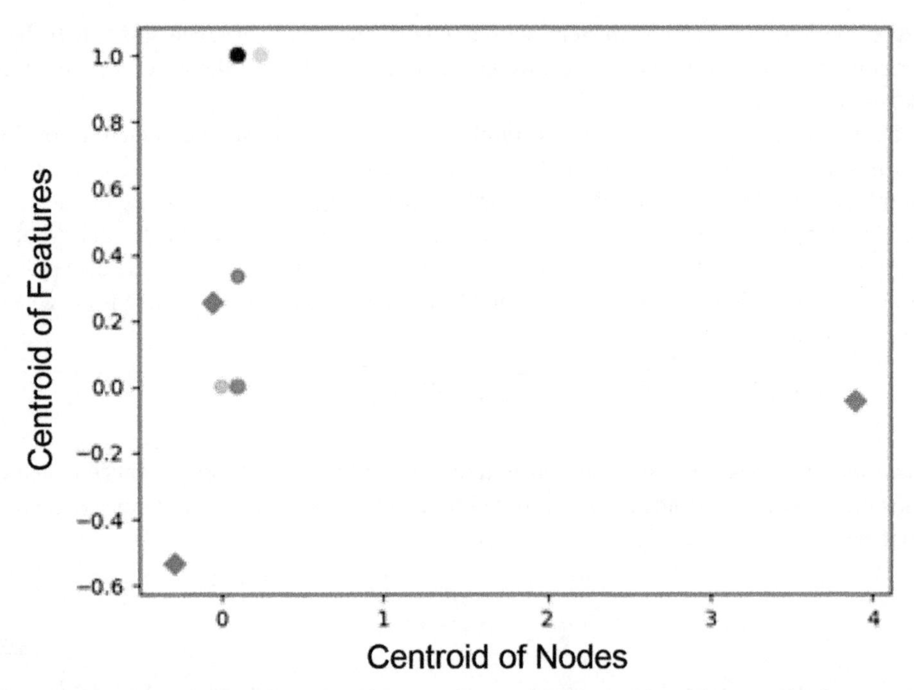

Figure 3.22 Anomaly detection using K-means cluster algorithm without feature selection.

shown in Figure 3.22. From the figure, it can be seen that K-means cluster algorithm has classified the total nodes into normal nodes and attacked nodes. The attacked nodes were also grouped into wormhole-attacked group and black-hole-attacked group. The performance metrics of the developed IDS is discussed in detail in the next chapter.

3.9.10 IDS Using K-Means Cluster Algorithm with Feature Selection

This K-means cluster classifier IDS incorporates feature selection through random forest method and PCA.

3.9.11 IDS Using K-Means Cluster Algorithm with Random Forest Feature Selection

By using random forest method, the following four features were selected for intrusion detection, from the 12 network layer features:

1) Route count changed in percentage.
2) Hop count changed in percentage.
3) Receiver node power changed in percentage.
4) Drop ratio.

With the help of the four features, the K-means cluster classifier identifies wormhole attacks in WANs. Figure 3.23 shows the K-means clusters of the features selected using random forest method. The clusters are named as Label 0, Label 1, and Label 2. The position of the three clusters is shown in Figure 3.23. The impact of the PRC, PHC, and drop

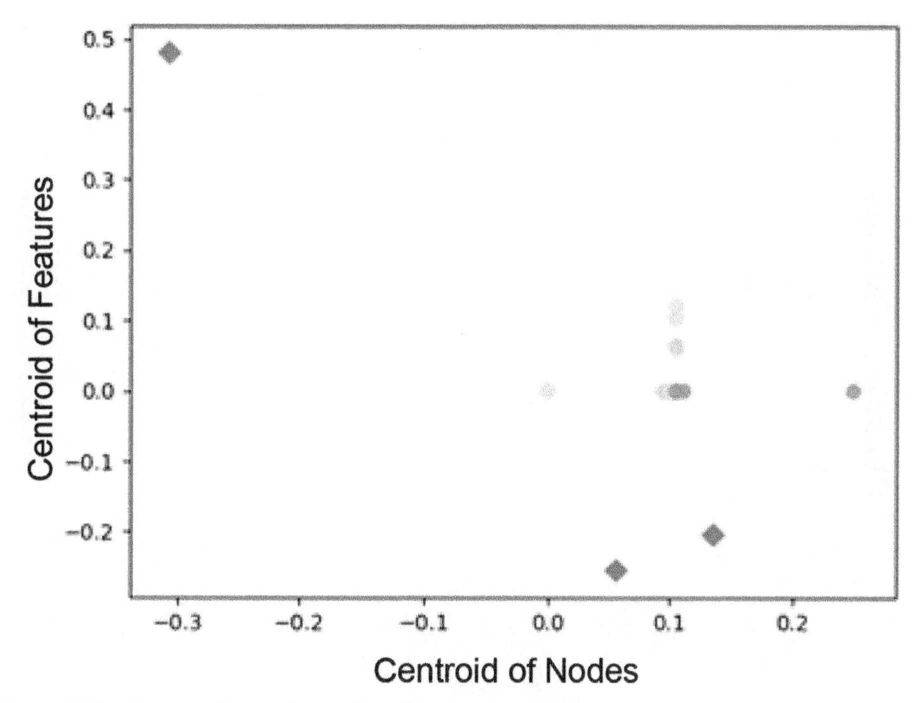

Figure 3.23 K-means clusters for random forest reduced data.

ratio is more in Label 0, whereas the impact of receiver power is more in Label 2. The nodes that fall on each label are noted down.

Figure 3.24 illustrates the cluster classification of the percentage-route-changed feature. The behavior of the abnormal node has a high variation in the percentage-route-changed parameter. Because of the high threshold of the percentage-route-changed feature value, the clusters shown beyond 0.6 fall away from the centroid and thereby exhibit an abnormal behavior.

Figure 3.25 depicts the position of clusters with respect to the percentage-hop-changed parameter. The clusters that fall above the threshold value of 200 are considered as abnormal behavior nodes. Similarly, the position of clusters with respect to the receiver-power feature is portrayed in Figure 3.26. The clusters that fall above the threshold value of 1 are considered as abnormal nodes.

3.9.12 IDS Using K-Means Cluster Algorithm with PCA Feature Selection

By using PCA, the following four features were shortlisted for intrusion detection from the 12 network layer features:

1) Average difference in the sequence number.
2) Maximum changes that occurred in hop count of the WAN.
3) Average difference in hop count.
4) Drop ratio.

Figure 3.27 shows the positions of clusters based on the packet-drop ratio. This packet-drop-ratio feature is one of the important parameters in WAN analysis.

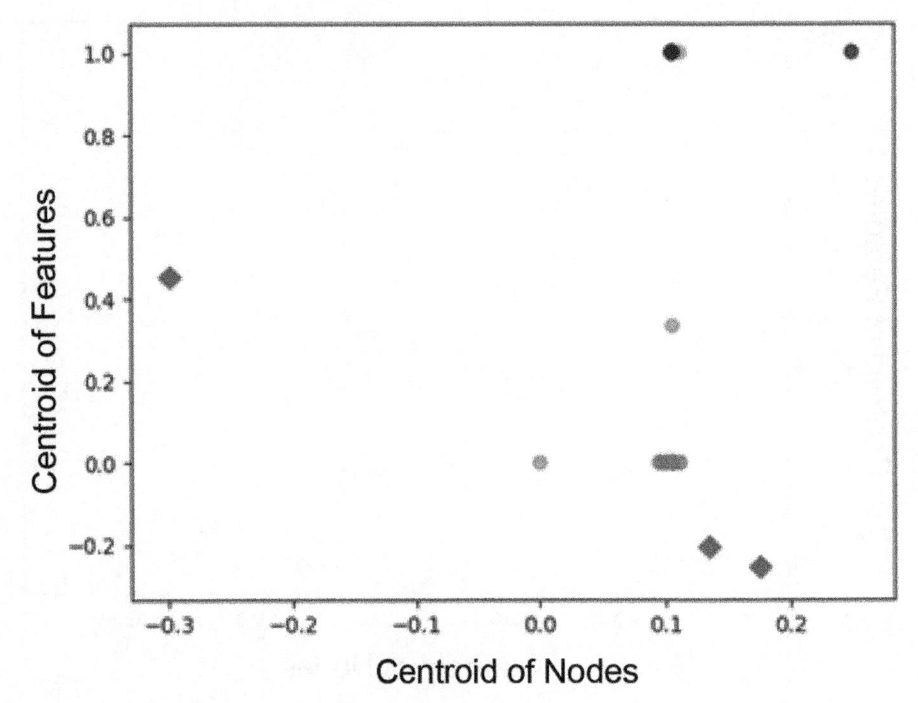

Figure 3.24 Cluster classification with percentage route changed.

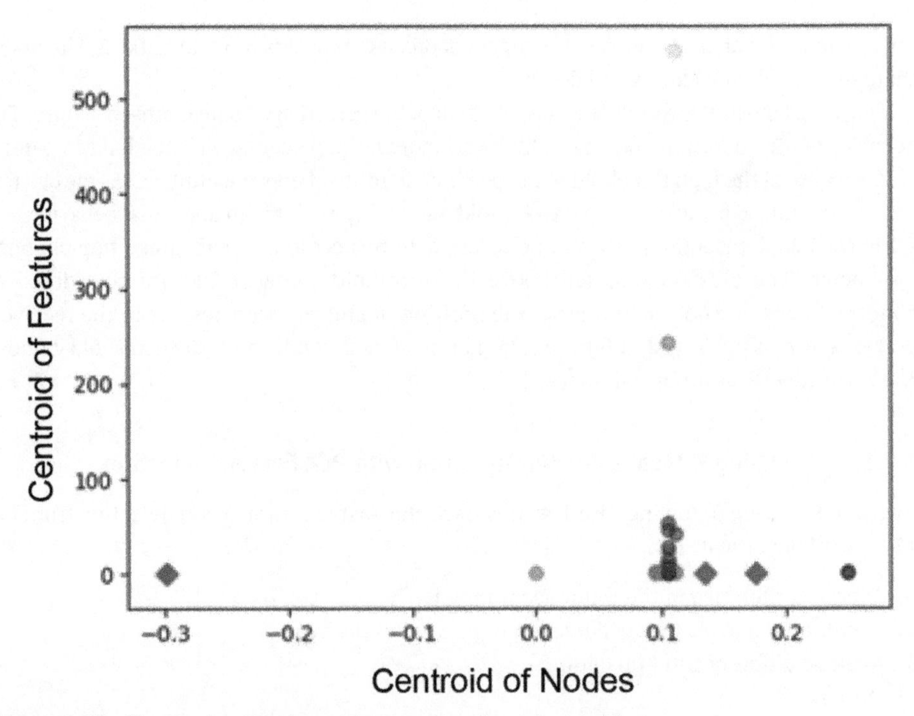

Figure 3.25 Cluster classification with percentage-hop-changed feature.

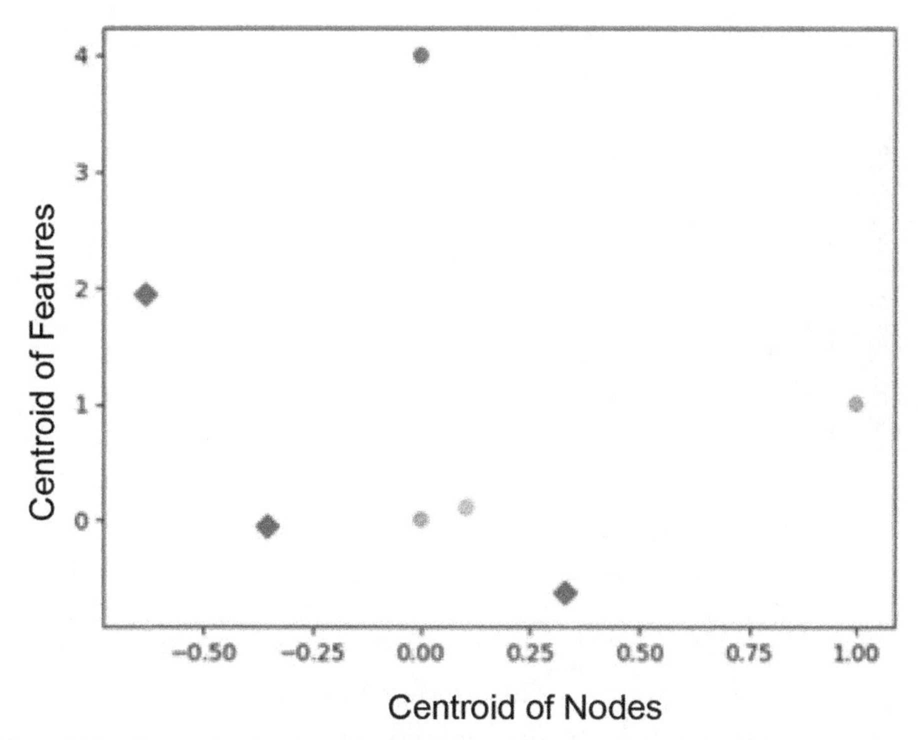

Figure 3.26 Cluster classification with receiver-power feature.

Figure 3.28 shows the cluster classification diagram for features selected through PCA. The clusters are determined using K-means cluster algorithm. Here, in Label 0, the impacts of Seq_diff_req_rep (average change in sequence number) and hop_diff_req_rep (average change in hop count) are more, whereas the influence of PHC and drop ratio (packet drop ratio) is more in Label 2. Using the abovementioned information, we can predict the type of attacks that happen in the network. For normal nodes, requisition rate, percentage of hop count changed, percentage of route change, and receiver power are within the threshold range. Based on the impact of features on clusters, a wormhole attack is identified. Wormhole attacks have a high packet drop ratio, more receiver power, and abnormal hopping behavior. Figure 3.29 shows the cluster position, based on the maximum changes in the hop count feature. In wormhole attacks, there is abnormality in the maximum-changes-in-the-hop-count feature. Figure 3.30 illustrates the position of clusters based on the percentage-hop-count-changed feature. This feature plays a vital role in the detection of wormhole attacks.

3.10 Results and Discussions

3.10.1 Performance Metrics of the IDS

Mostly, the performance metrics of IDS in WANs is computed using the following.

a) *Detection rate* gives the rate of correctly identified intrusions in the WAN from the total intrusions in the network:

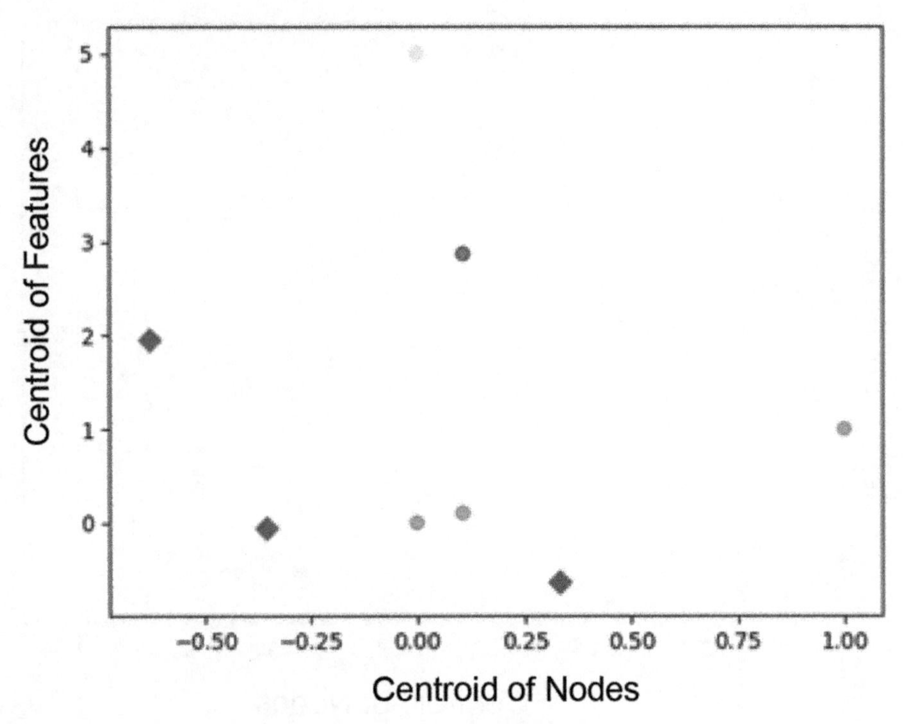

Figure 3.27 Cluster classification with packet-drop-ratio feature.

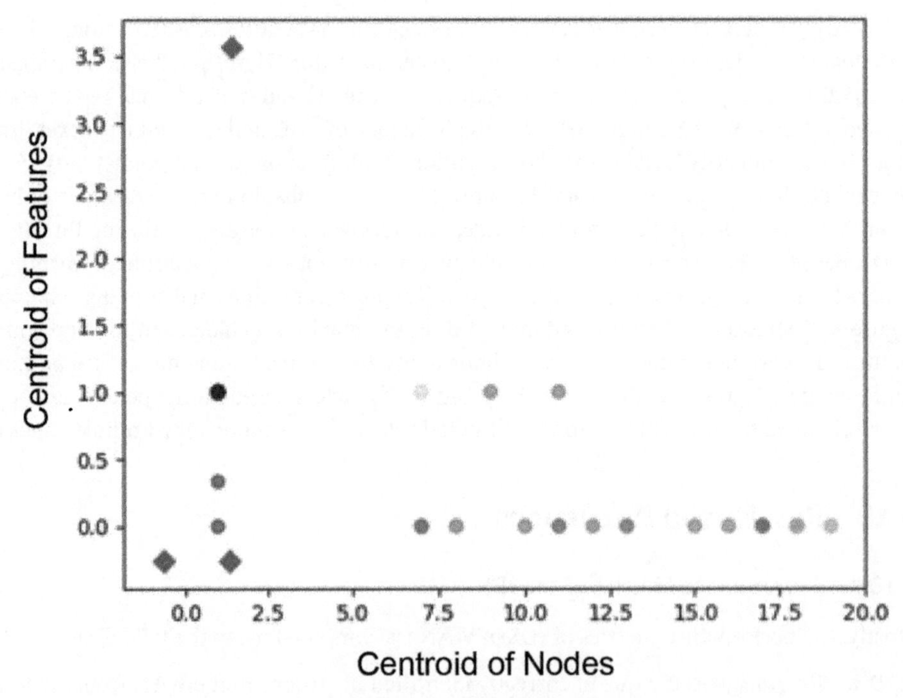

Figure 3.28 K-means cluster classification using PCA selected features.

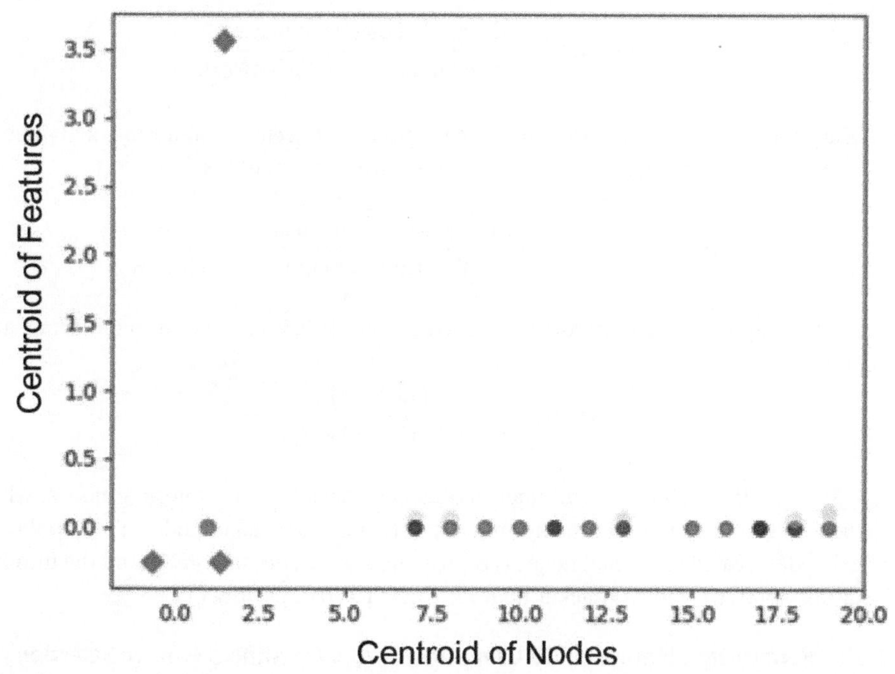

Figure 3.29 Cluster classification of PCA + K-means IDS with maximum changes in the hop count feature.

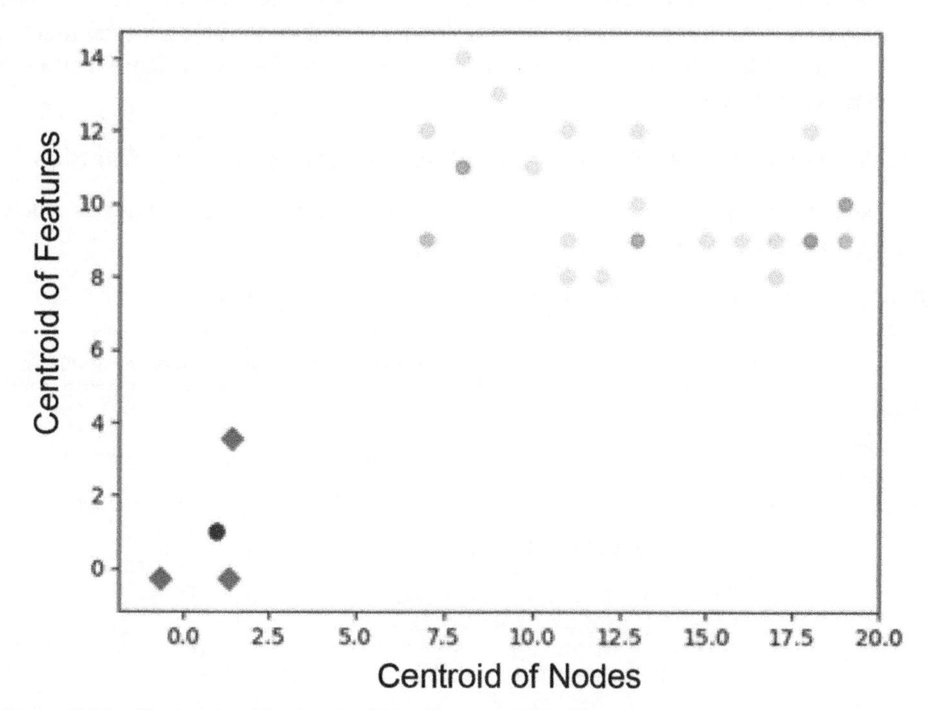

Figure 3.30 Cluster classification for PCA + K-means IDS with percentage hop count changed.

$$\text{Detection rate} = \frac{\text{Correctly detected intrusions}}{\text{Total intrusions in the network}}$$

b) *False positive rate* gives the rate of normal activities detected as abnormal activities in the WAN. For a high detection rate, the false alarm rate should be low:

$$\text{False positive rate} = \frac{\text{Normal marked as intrusions}}{\text{Total normal activities}} = \frac{\text{FP}}{\text{TN+FP}}$$

c) *Accuracy* represents the proportion of correctly classified instances into malicious and benign:

$$\text{Accuracy} = \frac{(\text{TP+TN})}{(\text{TP+TN+FP+FN})}$$

Here, *false negative* (FN) is the number of attacker nodes treated as normal nodes, while *false positive* (FP) is the number of normal nodes treated as attacker nodes. The number of normal nodes treated as normal nodes is represented as *true negative* (TN), and the number of attacker nodes treated as attacker nodes is treated as *true positive* (TP).

3.10.2 Performance Metrics of IDS Using One-Class SVM without Feature Selection

The performance metrics of one-class SVM (without feature selection) classifier used to build the IDS for the WAN is presented in Table 3.5.

For the one-class SVM IDSs with 12 network layer features, the detection rate is 99% and accuracy is 96.67%. The false positive rate is as low as 3.5%. A lower value indicates that the performance of the IDS is good. Figure 3.31 gives the plot for the performance metrics of one-class SVM classifier.

3.10.3 Performance Metrics of IDS Using One-Class SVM with Feature Selection

Twelve network layer features were reduced to four using the machine learning models, random forest method, and PCA techniques.

Table 3.5 Performance metrics of one-class SVM without feature selection.

IDS performance metrics	One-class SVM (without feature selection)
Detection rate	99%
Accuracy	96.67%
False positive rate	3.5%

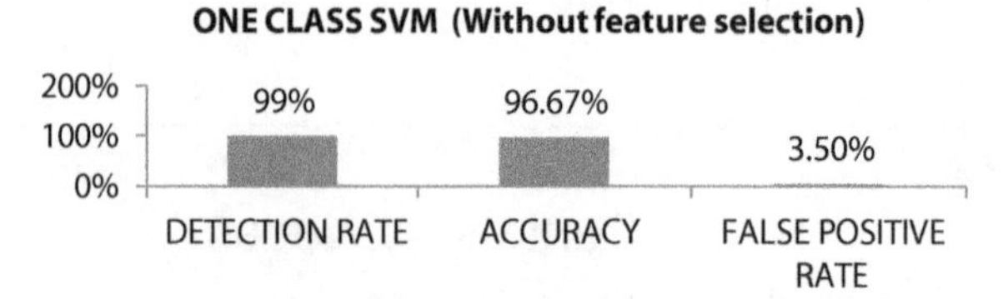

Figure 3.31 Anomaly detection in one-class SVM without feature selection.

The performance metrics of the IDS built using one-class SVM with random forest feature selection technique is given in Table 3.6. As can be seen, the accuracy of the system improved to 97.7%, and the FPR reduced to 2.35%, which is better than one-class SVM without feature selection.

The four network layer features were selected from the 12 network layer features using the PCA method. Using these four features, IDS was built. Here, the accuracy improved to 98.8%, and the FPR reduced to 1.17% than to the one-class SVM without feature selection IDS as presented in Figure 3.32. The performance metrics and graphical representation is presented in Table 3.7 and Figure 3.33.

Table 3.6 Performance metrics of the RF + one-class SVM IDS.

IDS performance metrics	RF + one-class SVM IDS
Detection rate	99%
Accuracy	97.7%
False positive rate	2.35%

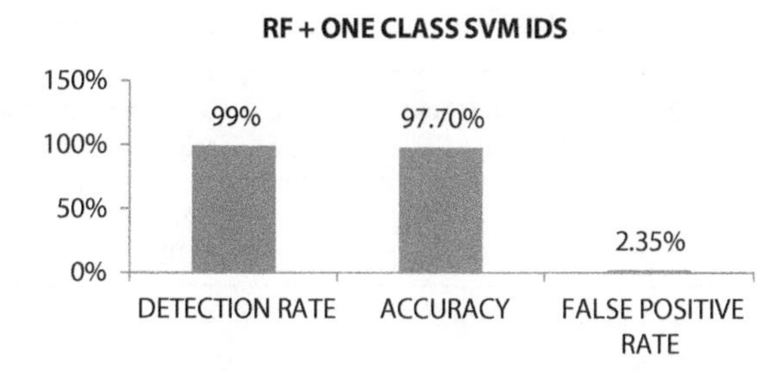

Figure 3.32 IDS metrics of RF + one-class SVM.

Table 3.7 Performance metrics of the PCA + one-class SVM IDS.

IDS performance metrics	PCA + one-class SVM
Detection rate	99%
Accuracy	98.8%
False positive rate	1.17%

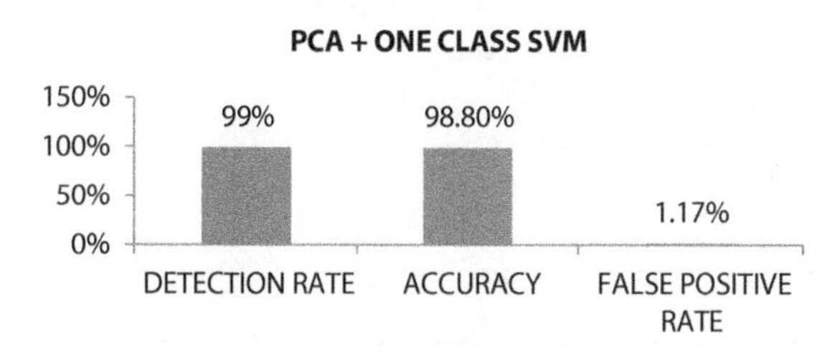

Figure 3.33 Performance metrics of PCA + one-class SVM IDS.

3.10.4 Comparison of Performance Metrics of IDS Using One-Class SVM with and without Feature Selection

The comparison of performance metrics of the IDS using one-class SVM with and without feature selection is given in Table 3.8. Figure 3.34 shows the graphical illustration. From this, it is inferred that, with feature selection, the accuracy and false detection improve. The false positive rate reduces randomly.

It is demonstrated that the IDS designed using the features selected through PCA method has very low false positive rate as compared to the IDS designed using the features selected through random forest method.

3.10.5 Detection of Performance Metrics of IDS Using K-Means Cluster Classifier without Feature Selection

The IDS performance metrics of the K-means cluster classifier with 12 network layer features are given in Table 3.9 and its performance metrics in Table 3.10. In the detection of wormhole attacks, the detection rate is 96%, accuracy is 92.2%, and false positive rate is 9.2%. It is represented in Figure 3.35.

Table 3.8 Comparison of one-class SVM with and without feature selection.

IDS performance metrics	One-class SVM (without feature selection)	One-class SVM (with feature selection)	
		RF + one-class SVM IDS	PCA + one-class SVM IDS
Detection rate	99%	99%	99%
Accuracy	96.67%	97.7%	98.8%
False positive rate	3.5%	2.35%	1.17%

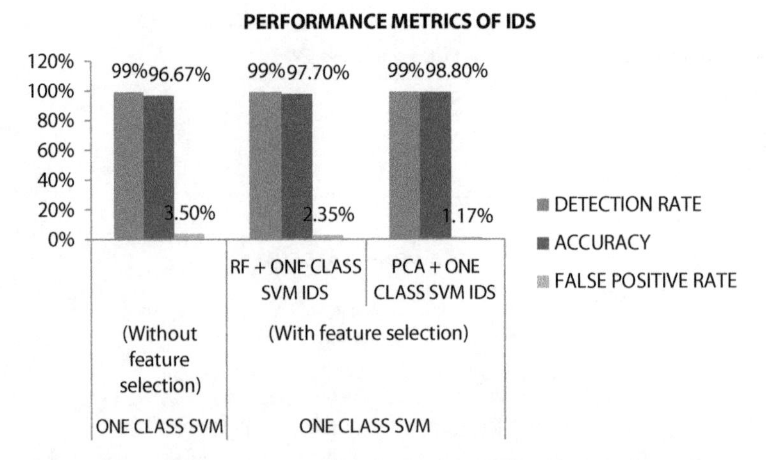

Figure 3.34 Comparison of performance metrics of one-class SVM with and without feature selection.

Table 3.9 Performance metrics of IDS using K-means cluster algorithm without feature selection.

IDS performance metrics	K-means-cluster-algorithm-based IDS without feature selection in wormhole attack detection
Detection rate	96%
Accuracy	92.2%
False positive rate	9.2%

Table 3.10 Performance metrics of K-means cluster classifier IDS with feature selection.

IDS performance metrics in wormhole detection	RF + K-means IDS	PCA + K-means IDS
Detection rate	99%	99%
Accuracy	97.70%	98.80%
False positive rate	2.30%	1.17%

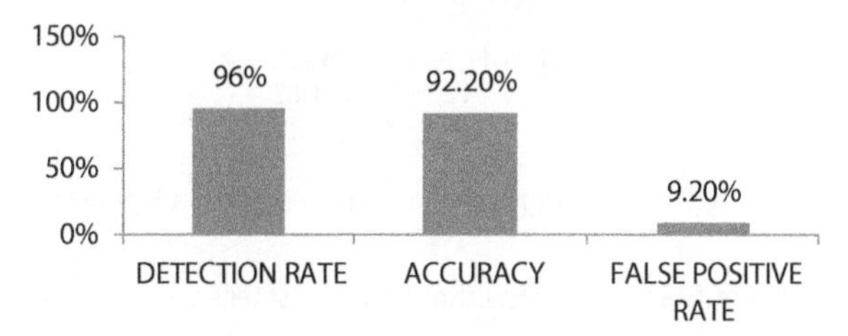

Figure 3.35 Performance metrics of IDS using K-means cluster algorithm without feature selection.

3.10.6 Performance Metrics of K-Means Cluster Classifier IDS with Feature Selection Technique

In the detection of wormhole attacks, the accuracy of the network is 97.7%, the detection rate is nearly equal to 99%, and the false positive rate is 2.3%. And, by using PCA feature selection technique, the detection rate is 99%, accuracy is 98.80%, and FPR is 1.17%. Among these IDSs, the PCA + K-means IDS performs well. The graphical representation for these IDS is given in Figure 3.36. The comparison of the IDS that is constructed using K-means cluster classifier with and without feature selection is given in Table 3.11, and its graphical representation is given in Figure 3.37. Among these systems, with feature selection IDS has a very low false positive rate and PCA + K-means IDS performance is good. When compared with conventional systems PCA integrated with K-means algorithm achieves higher performance.

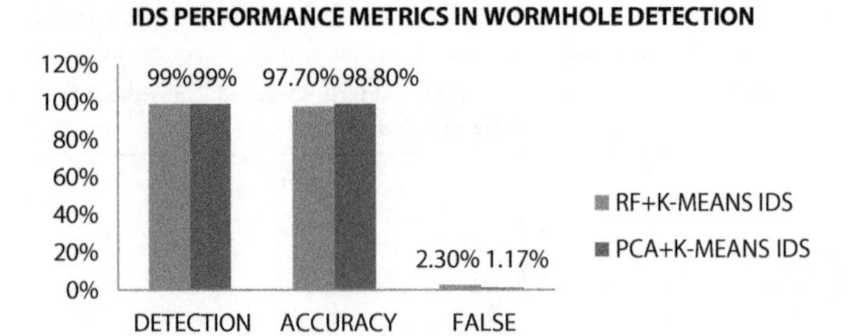

Figure 3.36 Performance metrics of K-means cluster classifier IDS with feature selection.

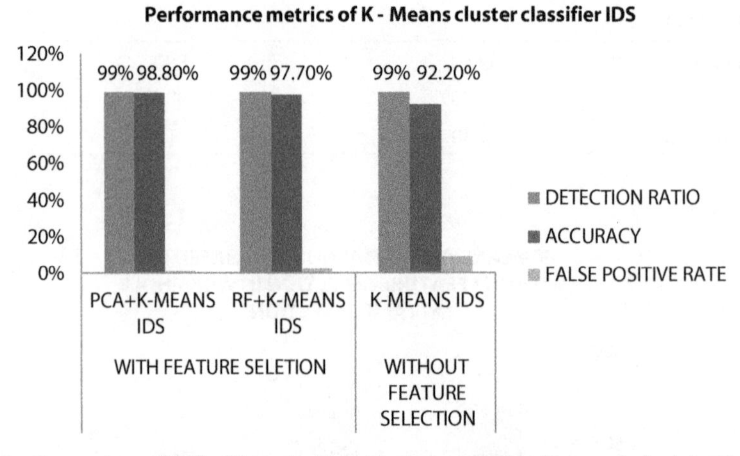

Figure 3.37 Comparison of IDS with and without feature selection in wormhole detection.

Table 3.11 Comparison of IDSs with and without feature selection in wormhole attack detection.

IDS performance metrics	With feature selection		Without feature selection
	PCA + K-means IDS	RF + K-means IDS	K-means IDS
Detection ratio	99%	99%	99%
Accuracy	98.8%	97.7%	92.2%
False positive rate	1.17%	2.3%	9.2%

3.10.7 Comparison of Performance Metrics of Designed IDS with Previously Existing IDS

The IDS designed in the present study is compared with previously existing IDSs designed by other investigators. Table 3.12 gives the comparison of the performance metrics of the existing IDS based on genetic-based fuzzy (GBF) and sequential pattern mining (SPM) in

Table 3.12 Comparison of existing and new IDS in the detection of wormhole attacks.

IDS techniques	Accuracy (%)	False positive rate (%)	Detection ratio (%)
(Existing) GBF%	96.8	14	95.5
(Existing) SPM %	88	48	80.6
PCA + K-means cluster %	98.8	1.17	100
RF + K-means cluster %	97.7	2.3	100

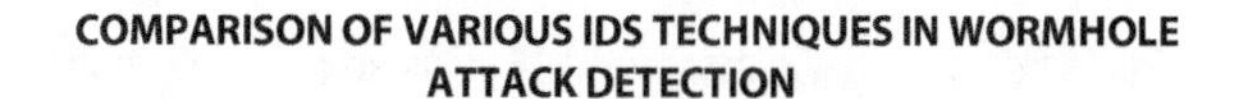

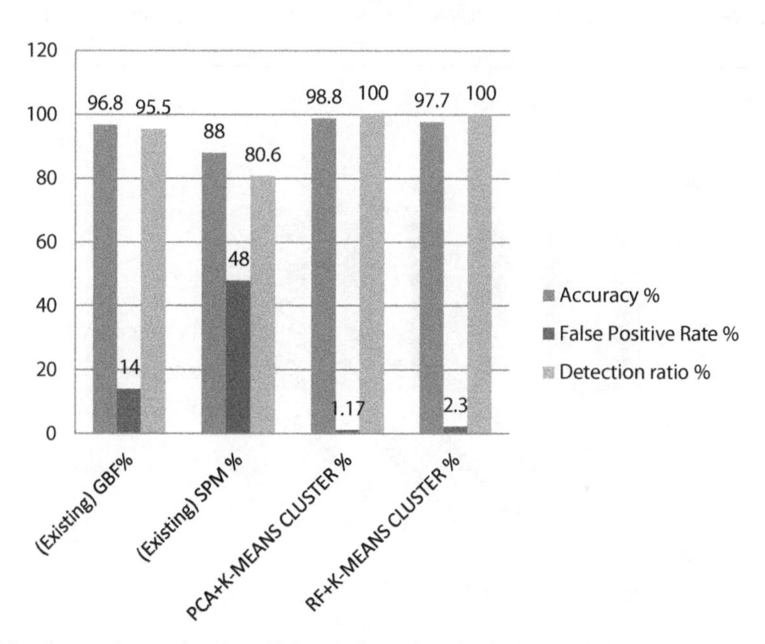

Figure 3.38 Comparisons of various IDS techniques in wormhole attack detection.

the detection of wormhole attacks. Figure 3.38 illustrates the corresponding comparison plot. It is clearly seen that the accuracy and detection ratio of the present study are higher than those in existing GBF- and SPM-based IDSs. The false positive rate of the designed IDS is much lower when compared to existing old IDSs.

3.10.8 Comparison of Presented IDS in Wormhole Attack Detection

In this proposed IDS, the wormhole attack detection system with feature selection performs well as compared to without feature selection. The IDS that is selected with PCA has a low false positive rate of 1.17% in the detection of wormhole attacks, as presented in Table 3.13 and Figure 3.39.

Table 3.13 Comparison of proposed IDS.

IDS parameter	One-class SVM without feature selection	K-means without feature selection in wormhole attack	With feature selection in wormhole attack		With feature selection in one-class SVM	
			PCA + K-means IDS	RF + K-means IDS	PCA + one-class SVM	RF + one-class SVM
Detection rate	99%	96%	99%	99%	99%	99%
Accuracy	96.67%	92.20%	98.80%	97.70%	98.80%	97.70%
False positive rate	3.50%	9.20%	1.17%	2.30%	1.17%	2.35%

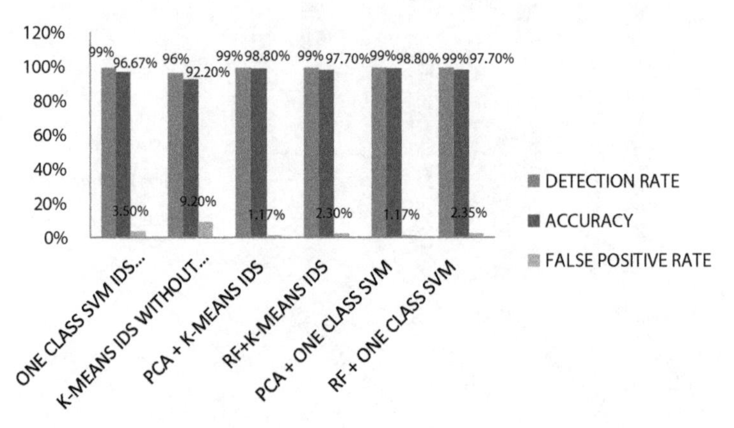

Figure 3.39 Comparison of proposed IDS.

3.11 Conclusion

IDSs play a vibrant role in identifying the threats encountered in a network by detecting changes in the normal profile due to attacks. Machine learning models such as one-class SVM and K-means cluster classifiers were used to build the IDSs.

An IDS was built using one-class SVM having 12 network layer features. The accuracy and detection rate of this IDS were 96.67% and 99%, respectively. The FPR was 3.5%. Although the accuracy and detection ratio were good, FPR could be improved. And with network layer feature selection, four features are selected from 12 features using random forest and PCA methods. Here, the FPR values are 1.17% and 2.35% when RF and PCA are used, respectively. It was evidently observed that the one-class SVM with PCA method for feature selection was the best when other two IDSs were designed using one-class SVM.

Similarly, while using K-means cluster classifier IDS, the accuracy and detection rates were 92.2% and 96%, respectively, and FPR was 9.2% in the detection of wormhole attacks without feature selection. In RF + K-means IDS, the accuracy and detection rates were 97.7% and 99%, respectively, and FPR was 2.3%. In PCA + K-means IDS, the accuracy and detection rates were 98.8% and 99%, respectively, and FPR was 1.17%.

The IDSs designed using PCA + K-means cluster classifier IDS is very effective in the detection of wormhole attacks, whereas PCA + one-class SVM IDS is good in the detection of wormhole attacks in WANs.

3.12 Scope for Future Work

The IDS designed in the present work can be used to detect the types of attacks that affect WANs. However, only one attack was detected in the present study; hence, further studies need to be undertaken to detect more attacks. Also, other machine learning algorithms could be tried to improve the detection ratio and accuracy by implementing new technologies for data preprocessing and feature selection. Furthermore, studies could be undertaken to minimize the processing time.

In the present work, the detection time taken by the system is 5 seconds.

References

[1] R. Bace and P. Mell, "Intrusion Detection Systems," 2001. http://csrc.nist.gov/publications/nistpubs/800-31/sp800-31.pdf

[2] Paulo angelo alves resende and andré costa drummond, "A survey of random forest based methods for intrusion detection systems," *ACM Comp. Sur.*, vol. 51, no. 3, Article 48, 2018.

[3] R. De Pietro and L.V. Mancini, Intrusion Detection System, 2008, Springer publications, 2008.

[4] J. Geier, Wireless Networks First-step, Cisco press, chapter 2, 2004.

[5] T. Jackson, J. Levine, J. Grizzard, and H. Owen, "An investigation of a compromised host on a honeynet being used to increase the security of a large enterprise network." In *Proceedings of the 2004 IEEE Workshop on Information Assurance and Security*, 2004; pp. 9–14.

[6] Y. Xiao, V.K. Rayi, B. Sun, X. Du, F. Hu, and M.A. Galloway, "survey of key management schemes in wireless sensor networks." In *Computer communications, Science direct*, Elsevier, 2007; vol. 30, pp. 2314–2341.

[7] Y. Zhou, Y. Fang, and Y. Zhang, "Security wireless sensor networks: A survey," *IEEE Comm. Surv.*, vol. 10, no. 3, pp. 6–28, 2008.

[8] R.E. Kaissi, A. Kayssi, A. Chehab, and Z. Dawy, "DAWWSEN: A defense mechanism against wormhole attack in wireless sensor network." *Proceedings of the Second International Conference on Innovations in Information Technology*, 2005.

[9] S. Devendra and S.S. Bedi, "A survey: Feature based intrusion detection system in mobile ad-hoc network," *Int. J. CST*, vol. 6, no. 4, pp. 135–140, 2015.

[10] S. Sen and Z. Dogmus, "Feature selection for detection of adhoc flooding attacks, advances in intelligent systems and computing: Preface." In *Advances in Computing and Information Technology*, 2012; pp. 11–22.
[11] P. Kabiri and M. Aghaei, "Feature analysis for intrusion detection in mobile ad-hoc networks," *Int. J. Net. Sec.*, vol. 12, no. 1, pp. 42–49, 2011.
[12] L. Hi, J. Xia, et al., "An efficient intrusion detection system based on support vector machines and gradually features removal method," *Exp. Syst. Appl.*, vol. 39, no. 1, pp. 424–430, 2012.
[13] A. Tesfahun and D.L. Bhaskari, "Intrusion detection using random forests classifier with SMOTE and feature reduction." In *International Conference on Cloud & Ubiquitous Computing & Emerging Technologies*, 2013; pp. 127–132.
[14] F.Q.Z. Wenying, "Mining network data for intrusion detection through combining SVMs with ant colony networks." In *Future Generation Computer Systems*, Elsevier, 2014; vol. 37, 127–140.
[15] S. Madria and J. Yin, "SeRWA: A secure routing protocol against wormhole attacks in sensor networks," *Ad Hoc Net.*, vol. 7, no. 6, pp. 1051–1063, 2009.
[16] H.S. Chiu and K.S. Lui, "DelPHI: Wormhole detection mechanism for ad hoc wireless networks." In *IEEE International symposium on wireless pervasive computing*, IEEE Xplore, 2006; pp. 2–12. doi:10.1109/iswpc.2006.1613586.
[17] H. Chen, W. Lou, Z. Wang, J. Wu, Z. Wang, and A. Xia, "Securing DV-Hop localization against wormhole attacks in wireless sensor networks," *Pervasive and Mobile Comp.*, vol. 16, pp. 22–35, 2015.

4

Neuro-Fuzzy-Based Bidirectional and Biobjective Reactive Routing Schema for Critical Wireless Sensor Networks

K.M. Karthick Raghunath[1] *and G.R. Anantha Raman*[2]

[1] *Associate Professor, MRIET, Secunderabad, Telangana*
[2] *Professor, MRIET, Secunderabad, Telangana*

4.1 Introduction

The challenges and risks of routing sensed data or newly generated events in a critical environment are always found to be a major research agenda for any researcher. A wireless sensor network (WSN) [1] deployed in critical infrastructure/environs mostly implies the difficulties to process the sensed data, which may raise problems in the compatibility factors of the network deployment—for example, jeopardizing scenarios and monitoring emergencies such as fire eruption and explosion and environment-oriented and hazardous pollution. In the future, one can expect that the monitoring of critical areas that are outfitted with a set of smart sensors will adopt some smart processing methodologies. These smart methodologies enhance the data transmission capabilities of the nodes deployed in the network. Moreover, the essential part of any critical environment is to deliver the sensed data as quickly as possible. Thus, routing plays a crucial role in the fast transmission of data and guarantees the quality functioning of the network in terms of reliability, timeliness, etc. Minimizing the delay factors in data transmission not only enhances the quality of service (QoS) of the entire network but also makes the actuator perform the corresponding actions in a limited time span, which is the primary objective of critical WSNs.

Routing in a dynamic network is quite challenging because some of the basic criteria of routing path establishment depend on hops, ways of communication, route requests and replies, unreachable destinations, speed of mobile nodes, etc. Especially in reactive routing strategies (where the construction of routing path begins only after the demand), path establishment and data transmission have to be dealt with narrowly. Failure to do so will lead to a deeply negative impact on the performance of the network. Nevertheless, high mobility in the network could disallow many essential transmissions to successfully route the sensed data [2]. But then, a more complicated routing protocol could demand extra stability considerations for deployed sensors to convey the information with the sink [3]. Despite the additional complications in designing the routing protocols for dynamic WSN

Sensor Data Analysis and Management: The Role of Deep Learning, First Edition. Edited by A. Suresh, R. Udendhran, and M.S. Irfan Ahmed.
© 2021 John Wiley & Sons, Ltd. Published 2021 by John Wiley & Sons, Ltd.

[1], the mobility concept finds an opportunity to reduce the involvement of the number of intermediate nodes in data transmission, which further minimizes the latency. Instant selection of randomized path with an idea of multiple options toward the destination along with minimized overheads can bring down the unbalancing effects on the network system, and this act may, to some extent, invoke the tolerating capabilities of nodes against possible failures. To provide such instant path selective options, especially in dynamic networks, it is essential to employ an advanced inference system.

One such advanced inference system of the current trend is the neuro-fuzzy system (NFS), which is the fusion technique of both fuzzy logic and neural network. Here, the system implies the core characteristics of fuzzy sets and the rule base, which are tuned at the adjustable (hidden) layers of the neural network iteratively using data vectors. Thus, the system exhibits the learning behavior in the first phase like the neural network, and the later phase demonstrates the behavior of fuzzy logic for further interpretation, execution, and system representation in a lucid state. This ultimately leads to the thought of enforcing the computational features of learning in fuzzy systems. Traditionally, the neural networks that comprise the set of efficient learning algorithms were constituted as an option to automatize or to aid in the fine-tuning of fuzzy rules. The term *neuro-fuzzy* was first quoted and investigated in the early 1990s by Lin, Jang, and Lee. Initially, it was focused mainly on control strategies, but, later on, the concept has been widely utilized in many applications such as data evaluation and classifications. Most of the neuro-fuzzy applications that evolved were of the process control type. The ways of utilization of NFS differ as per the applicability of input, weight, and output attributes for different application developments. Thus, Table 4.1 represents the various procedural types of NFS for better clarification and utilization in application development.

For handling critical data in dynamic WSNs, an intelligent routing strategy is always essential to enhance fast data deliverance and QoS in the network. In recent days, there are so many research proposals available regarding the optimal selection of route nodes from available intermediate nodes. Most of the research work for dynamic networks is location-centric, which raises some valid complications such as position inaccuracy and local

Table 4.1 Various types of NFS.

Type no.	Inputs	Weights	Output	Applications
0	Crisp	Crisp	Crisp	N/A
1	Fuzzy	Crisp	Crisp	Classification
2	Fuzzy	Crisp	Fuzzy	IF–THEN (Fuzzy)
3	Fuzzy	Fuzzy	Fuzzy	IF–THEN (Fuzzy)
4	Crisp	Fuzzy	Fuzzy	IF–THEN (Fuzzy)
5	Crisp	Crisp	Fuzzy	N/A
6	Crisp	Fuzzy	Crisp	N/A
7	Fuzzy	Fuzzy	Crisp	N/A

minimum [4]. Thus, among those existing systems, NFBBRR is instigated as a major proposed portion of this chapter. Further, the proposed NFBBRR is evaluated through repeated conduction of simulation, especially for determining an optimal route node in the dynamic and critical WSN. Finally, the resultant of the proposed system proved to be better than the other recent neuro-fuzzy and fuzzy systems. The performance ratios are found to be in the desired range under the unstable condition of the nodes in the network. The entire performance analysis has been clarified and justified in Section 4.7.

Section 4.1 furnishes a broader view on critical WSN, the necessity of smart routing, merits of NFS, and reasonable points to employ NFBBRR routing strategy in dynamic WSN. In Section 4.2, the working mechanism of recent research work is discussed to demonstrate the conventional procedural culture carried out through routing protocol for dynamic and critical WSNs. The key point of the proposed objective is illustrated in Section 4.3. The core theme of the proposed work and its contributions are delivered in Sections 4.4–4.6. As a perfect consequence to the proposed work presentment, Section 4.7 exposes the clarification on performance analysis. Finally, the chapter's conclusion part is summarized in Section 4.8, with some meritable points to support future enhancements.

4.2 Elemental Discussion: State-of-the-Art Review

4.2.1 Selection of Routing Node

Neuro-fuzzy is the combinatory technique of both fuzzy logic and neural network. The basic idea of neuro-fuzzy systems is to train the system (agent/network/subsystem) to simulate human decision-making capabilities, which further tends to solve the most complex problems under uncertainties. Such a system has the potential to deal with uncertain critical conditions and thus enhances the performance and self-reliant quality of the systems.

From Reference [5], the authors nominated a refined strategy that identifies the multipaths from the source node to the sink node. In this proposal, neuro-fuzzy methodology is utilized to determine the nullified node in WSN. Moreover, its ideology directs the source node to determine the routing path in the absence of any connectivity holes. The most optimal routing path is chosen among all possible routes from the source to the sink. The routing paths that are identified can sometimes be node-disorganized, link-disorganized, or overlapped with one another. This invalidates the paths that comprise multiple holes in the communication path.

Besides the core agendas, the proposal ensures the security of transmission data through encrypted dynamic confidential keys. The determined optimal path is confirmed to be the shortest routing and energy-efficient path. This reduces overall delay. However, the work fails to concentrate on the criticality of the data while rejecting routing paths with holes.

The proposed article in Reference [6] suggested a routing protocol for wireless mesh networks that employs the shortest routing path toward the sink. The sink node that is in the acquisition of generated data broadcasts a request message to the source node. On receiving the request, the source nodes acknowledge by sending a message to the sink and transmit the required information. Once the message has been completely transmitted, the source node gets the acknowledgment from the receiver side (hop/intermediate node), which is further carried on until the message reaches the sink. The acknowledging node is considered as routing nodes. Though the neuro-fuzzy method determines the optimal routing node (ORN), it fails to reduce the overall transmission delay.

In Reference [7], a new protocol based on an adaptive neuro-fuzzy inference system (ANFIS) was proposed for multipath construction using the factorial experimental design model (FEDM). Here, the datasets are generated and conveyed through FEDM. The dataset is gathered for critical review from the users, and the possible route path is ranked by the users based on the score. Around 162 data are utilized out of all possible datasets. The raw inputs are fuzzified through the fuzzification process concerning "high" and "low" limitations. The output membership function is represented with only one neuron in the output layer along with compliance/delay threshold rate representation.

The ANFIS infers the dataset for route utility via the training and checking process. Between the ranges of the input parameters, the membership functions and framed rule bases are fine-tuned by the inferring system. On the utilization of the back-propagation algorithm, the identified errors among the attained datasets are reduced through the selection of a desirable amount of epochs and fault-tolerance.

4.2.2 Mobility-Based Selection of Routing Node

The choice of a relaying node in the dynamic network is an essential job because it demands optimal speed, minimal energy, and predictable direction of the mobility of the nodes in the network.

The research article in Reference [8] projected a conceptualization that enables the node (that is nearer to the source) to conserve the minimum energy. This is because the chosen nodes that are nearer to the sink are always presumed to conserve minimum energy to route the data packets. Here, the solution is derived on proposing a kind of network with the essential mobile sink. Here, all the sink nodes are allowed to move simultaneously with the same acceleration in all possible directions. The performance of the stable network with the static sink nodes and an unstable network with the mobile sink nodes is analyzed. Here, the traffic load of the entire network gets reduced as an eventual process of employing mobile sinks, which further reduces the occurrence of congestion within the network.

The authors in Reference [9] proposed a new methodology that demonstrates an effective strategy for controlled mobility. This strategy instantly enhances the life expectancy (LE) of the network. The sink node moves across the network to gather the required information from the source nodes. To determine the motion and to draw the moving pattern of the destination node, a mixed integer linear programming (MILP) analytical model was applied. MILP delivers the solution to determine the set of possible routes toward the sink. Besides this, a greedy maximum residual energy method was also employed to move the destination node from its current position to a new position in such a way that the surrounding nodes possess the highest residual energy.

From Reference [10], it has been observed that the proposed four features of the mobility patterns for the destination node reduce the involvement of intermediate nodes to collect the required data through analysis strategies. Initially, the destination node tends to follow up a random walk mobility model for movement and the collection of data through passive fashion. As the destination node broadcasts a beacon signal periodically, each deployed node receives the beacon and manages to evolve the channel to carry the sensed information. Sometimes, this activity may lead to the collision of data. Second, in the proposed method, the sink tends to locomote on utilizing the concept of a random walk with restricted multihop propagation. The whole network area is segregated into square-shaped equal regions, and the centerfield of the region is viewed as a vertex. Then, the centers of each partitioned region are linked and thus frame the route for the sink node to move along with those center points. Hence, this process facilitates each node of the corresponding region to transmit the required message to the respective sink. Third, the sink node utilizes the biased random walk in gathering the data in a passive fashion. In the fourth procedure, for data propagation, multihop deterministic walk is used. Finally, all these procedures fail to ensure fast data deliverance to the sink node.

The method proposed in Reference [11] enables the mobility factor in wireless sensor nodes, especially for unmanned aerial vehicles. Here, the nodes tend to follow constant mobility speed in a homogeneous manner. Hence, this process is responsible for the alteration of network topology. Moreover, it poses the potentiality to detect critical events that emerge in uncertain regions.

The nominated mobility and clustering-based protocol in Reference [12] is particularly applicable to dynamic WSNs. Initially, based on residual energy, the sensor nodes choose itself as the supervisor node of the cluster based on parameters like the remaining energy and speed of the mobile node. A nonsupervisor node intends the link stability with a supervisor node, especially During the clustering process, a nonsupervisor node maintains link stability with a supervisor node based on the estimated connection timeout. Each noncluster head node is based on time division multiple access; a separate time slot is allocated to each supervisor node for data transmission through which the transmission schedule is carried out in ascending order. At a particular steady state, the transmission of sensed data and a request for joining a new cluster are accomplished in their allocated time slot. Thus, the process prevents enormous packet loss to the maximum extent, especially when the node joins with a new cluster.

4.2.3 Various Data Delivery Models in WSN

Several data delivery models employed in dynamic WSNs that are applied in exceptional applications are diagrammatically stated in Figure 4.1. Depending upon QoS demands in diverse applications, three data delivery models are constituted in the literature review. They are as follows:

- Query-driven model.
- Event-driven model.
- Continuous model [13, 14].

Query-driven model: Query-specific, delay-tolerant, time-critical, and interactive applications are categorized in this model. Here, queries play a primary role in driving systematical procedures. Queries can be sent in a reactive fashion, which further tends to minimize energy consumption. Usually, the base station fetches the data through queries when the

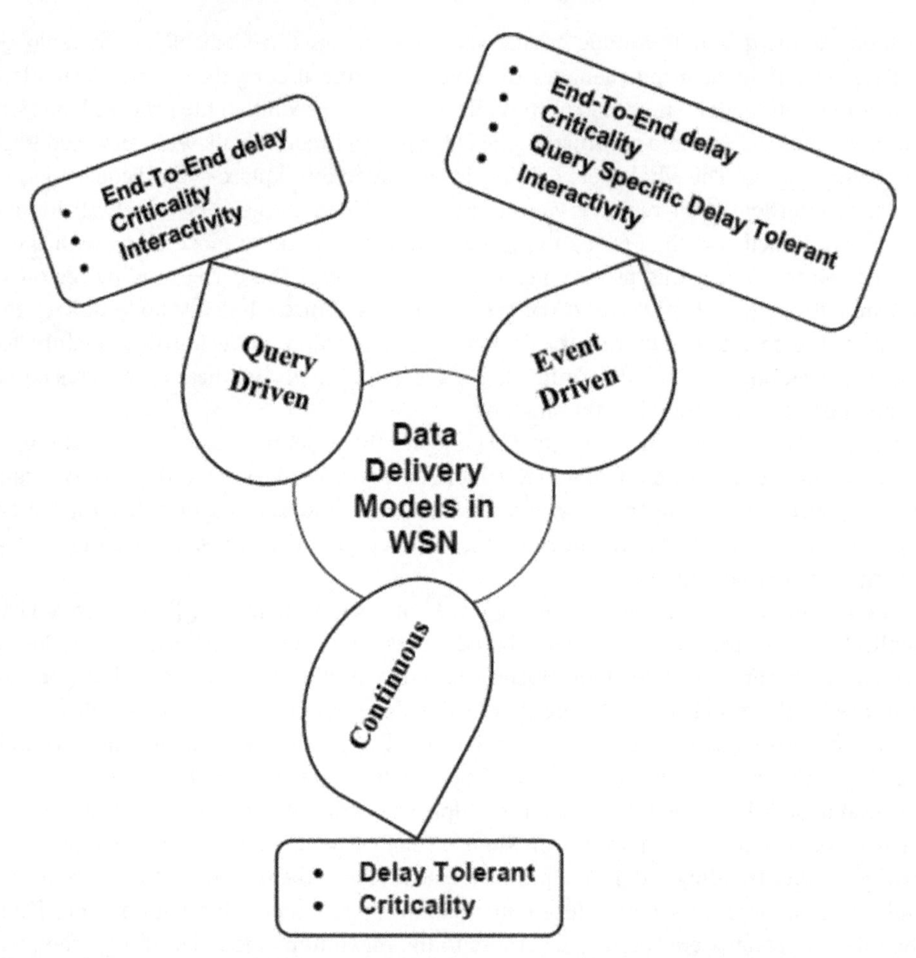

Figure 4.1 A requirement of data delivery models for various WSN applications.

deployed node generates essential on-demand events. Some examples of query-driven, model-based WSN applications are habitat monitoring and environmental surveillance/controlling systems.

Event-driven model: Mission-critical, delay-intolerant, and interactive applications are categorized in this model. Most of the event-driven, model-based applications are utilized as preventive countermeasures for the detected/concluded result to resolve and make the correct decisions as early as possible. Effective event spotting and precise deliverance of notification are some useful features of these application models.

Continuous model: Irrespective of demand, this model processes continuously through proactive mode for gathering the sensed data and reporting it to the end-user side.

The range of application necessities based on these three models is unlike in terms of various features. On utilizing these kinds of diverse models, most of the application-specific desired practicalities can be attained in WSNs.

4.3 Biobjectives of NFBBRR Model

This section presents a biobjective conceptualization. Here, choosing an optimal routing node and enabling fast data deliverance based on the opportunistic neuro-fuzzy inferring method are considered as biobjectives, which are further discussed in Sections 4.3.1 and 4.3.2.

4.3.1 Identification of an Optimal Routing Node

Optimal routing node determination has always been found to be one of the significant causes of concern in dynamic WSN due to its unstable infrastructure [15–17]. Furthermore, the prediction of quick and reliable routing paths has vast importance, especially for unstable WSN infrastructure where the routing path is highly differential, nodes are mobile, energy is limited, signal strength of different nodes is varying, and the sensor nodes have no fixed position. Thus, the identification of ORN is preferred as one of the objectives of the NFBBRR model.

4.3.2 Fast Route Establishment in Dynamic Networks

It is generally difficult to manage any routing strategy in dynamic networks. Hence, it is wise to establish a fast routing path based on opportunistic facts. Once the construction of the routing path is accomplished, most processes of data transmissions are carried out smoothly in the critical environment. In the proposed system, quick formation and organization of the routing path are preferred as other prominent objectives and managed on focusing bidirectional routing strategy, which is briefly discussed in Section 4.5.

4.4 Neuro-Fuzzy-Based Bidirectional and Biobjective Reactive Routing Schema

Considering the necessity for an efficient routing protocol in critical WSNs, this chapter evolves a novel routing protocol, *neuro-fuzzy-based bidirectional and biobjective reactive routing schema* (NFBBRR), which enables fast route establishment and ensures fast data deliverance from mobile nodes to the sink. NFBBRR utilizes a neuro-fuzzy concept for determining an optimal routing node to relay normal or critical events toward the targeted node. To ensure fast data delivery, the constitution of the routing path is optimized on employing bidirectional path establishment strategy. The working mechanism involved in the identification of optimal routing node and bidirectional route establishment is elaborated in Section 4.4.

4.4.1 NFBBRR Model

Improvising the overall QoS of the network is also considered to be a major issue that has to be dealt with cautiously. The NFBBRR model comprises the sequence of a process for the successful transmission of data (normal or critical) from any mobile node to the sink. Moreover, the proposed model is of reactive protocol type. Figure 4.2 depicts the sequence diagram of the NFBBRR model. Initially, the sink node in the network broadcasts the beacon signal as a requisition of newly generated events.

On obtaining the broadcasting event requisition beacon signal, the mobile node that is ready to report the detected event sends an event notification beacon signal to the sink. Further, the sink starts to establish the routing path toward the acknowledged mobile (target) node. Simultaneously, the target node that is responsible for transmitting the event information starts to find a routing path toward the requested sink node. The packet structures of both event requisition and event notification are depicted in Figures 4.3 and 4.4, respectively. The path establishment is carried out from both ends, that is, from the source to the sink and simultaneously from the sink to the source, using the opportunistic neuro-fuzzy mechanism (ONFM). ONFM is a new system in routing for dynamic and critical WSN infrastructure where relay nodes are chosen from among the closest and most efficient neighbor nodes for routing data toward the target node. For opting for the best routing node, it necessitates all the neighbor nodes in the network to exchange some essential parameters. The obtained result after the implementation of ONFM shows that it has enhanced the efficiency, QoS, and reliability of the critical WSNs.

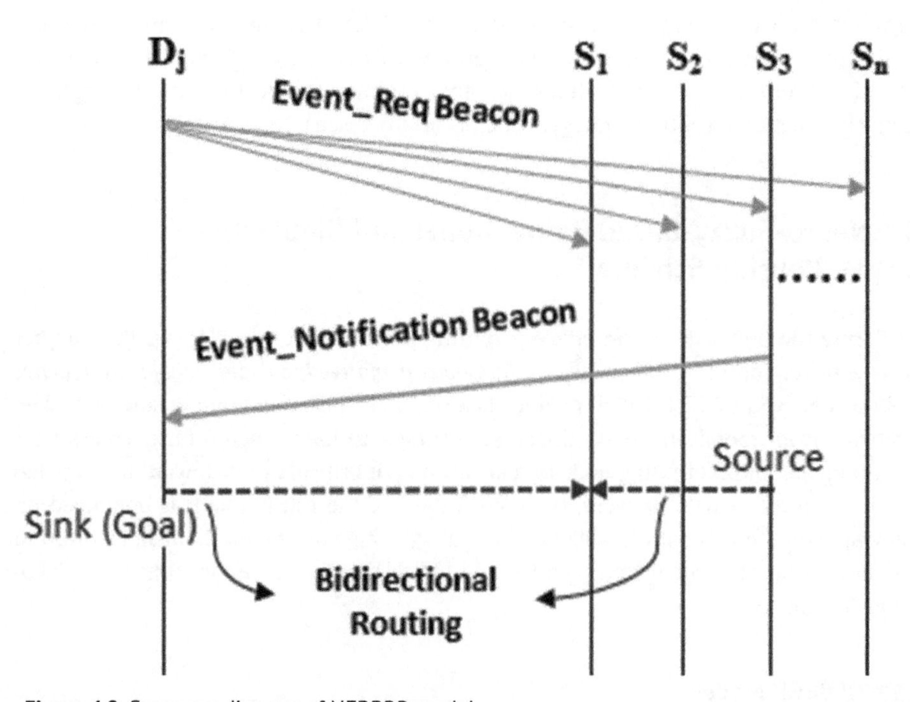

Figure 4.2 Sequence diagram of NFBBRR model.

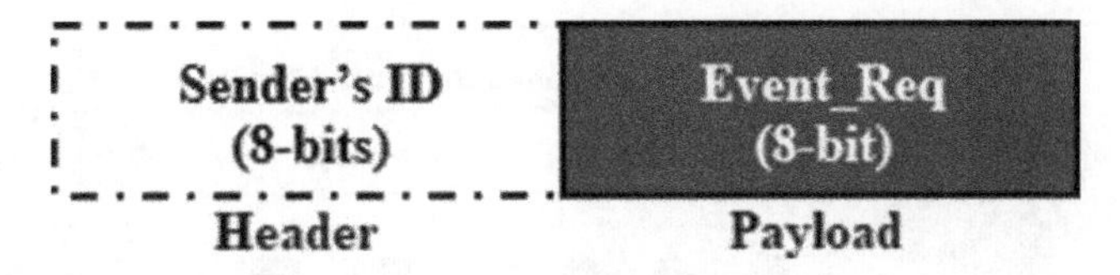

Figure 4.3 Packet structure of event requisition.

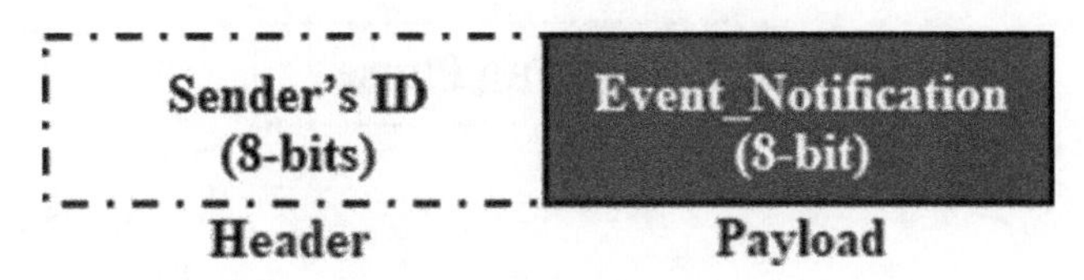

Figure 4.4 Packet structure of event notification.

4.4.2 Routing Phase

Apart from the routing control overheads, this section discusses the route construction and data transmission procedures of the NFBBRR model. Before the commencement of route establishments for each event reporting, all the neighbor nodes tend to exchange information like the current energy level (EL) of each node, current life expectancy of each node, and the average node stability time (NST). The packet structure of the neighbor node's information is depicted in Figure 4.5.

Here, the sink node tries to construct the route path toward the source node by propagating through its neighbor nodes; similarly, the source node tries to form the path through its neighbor nodes based on the exchanged information. Hence, all this exchanged information is prominently needed as input parameters to the ONFM. This mechanism portrayed in Figure 4.6 is composed of a powerful inference system and rule manager with rule base, which aids in determining the optimal intermediate routing node for sink and source. Further, the most elaborated process of ONFM has been discussed in Section 4.6.

4.5 Bidirectional Routing Strategy

A bidirectional routing strategy is a routing technique that constructs the routing path in two ways: from forward route establishment (from source to sink) and backward route establishment (from sink to source). The two-way construction commences from both sink and source simultaneously and terminates at a route point. The *route point* is defined as the point where two opposite directional constructions intersect, and this intersectional point commonly refers to one of the intermediate nodes in the network. For proceeding from both sink and source, the breadth-first search (BFS) [18] technique has been utilized, which gives completeness to the routing strategy. To apply BFS, all the one-hop communication nodes from the sink as well as from source are accounted to be in level_1, whereas

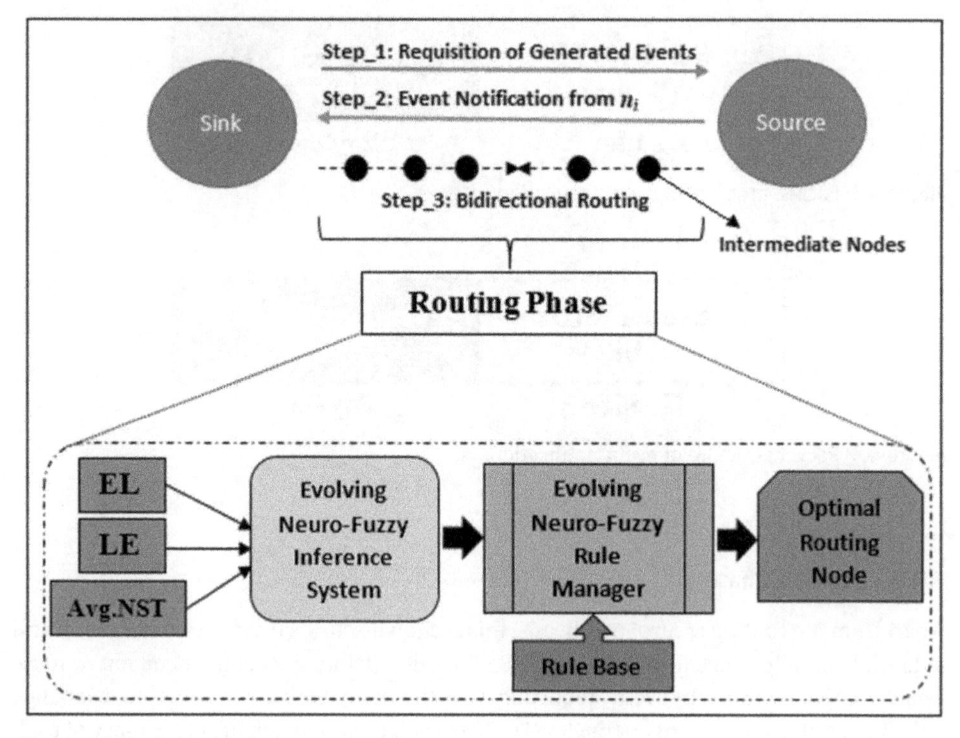

Figure 4.5 Overall architecture of NFBBRR model.

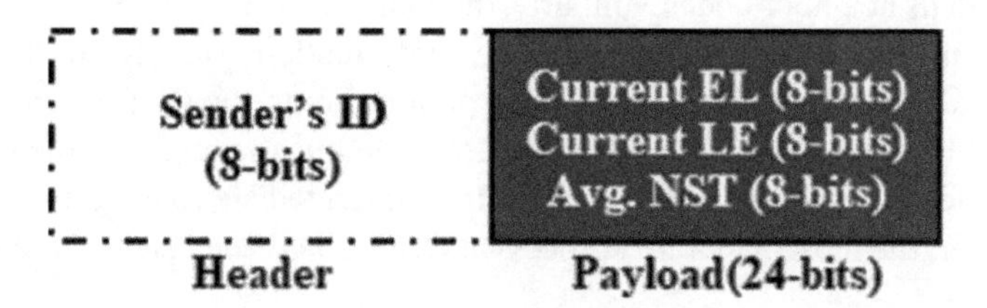

Figure 4.6 Packet structure of neighbor node's information.

all the two-hop communication nodes are accounted to be in level_2, and so on. Figure 4.7 depicts the generic overview of dynamic WSN for BSF-based bidirectional routing strategy.

4.5.1 BFS-Based Bidirectional Scenario

On utilizing the BFS traversing concept, both sink and source nodes start from a node n_i and mark it as tagged (regardless of target node). Initially, during the route establishment, the node n_i is tagged as unexplored. The node n_i is tagged as explored only when it is completely traversed in the motive of attaining the target node either by the sink or source node. All the remaining untraversed adjacent n_{i+1} nodes from the current n_i are traversed.

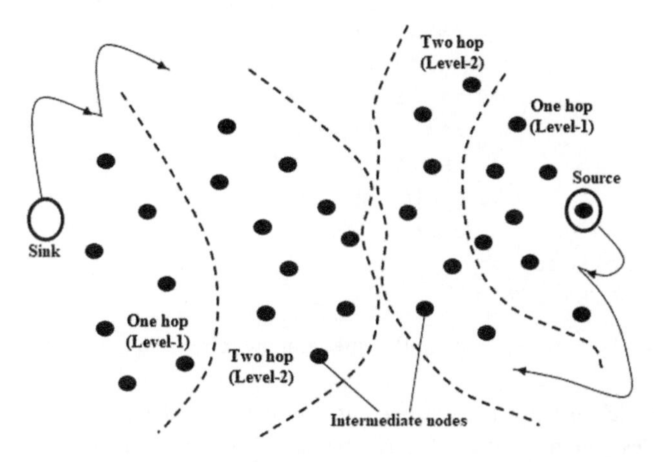

Figure 4.7 Generic view of dynamic WSN for BSF-based bidirectional routing strategy.

The newly traversed n_i which is not yet tagged is put on to the end of a list of untraversed n_i. The first n_i that is to be traversed is the next to be tagged on this queue list, and the same process is repeated until all the n_i are completely traversed and tagged at *level*$_i$. The n_i which are updated in a queue list are supposed to be traversed and tagged one by one, and, similarly, their corresponding neighboring n_i in next *level*$_i$ are also traversed and tagged.

Sometimes, an empty queue list may indicate the termination state of route establishment. All the states of n_i are qualified based on the current n_i state of the node. The major states of n_i are prepare state, awaiting state, and action state. The prepare state specifies the set of n_i which are yet to be traversed. Before the commencement of the entire process, all n_i are assumed to be in prepare state.

Algorithm I: BFS-Based Bidirectional Routing Strategy

```
qi. infix(xi) ^ tag xi ← dealt
qg. infix(xg) ^ tag xg ← dealt
        while qi ≠ 0 ^ qg ≠ 0 do
        if qi ≠ 0
            x ← qi.getFirst( )
            if x = xg ˅ x ∈ qg
                  return ATTAINED
            ∀[k∈ K(x)]
                  x'← f (x, k)
                  if x'← !dealt
                  tag x'← dealt
                  qi.infix (x')
                  else
                       conclude duplicate x'
            if qg≠ 0
            x' ← qg.getFirst( )
            if x' = xi ˅ x' ∈ qi
```

```
                return ATTAINED
        "[k⁻¹∊ K⁻¹(x')]
                x ← f⁻¹(x', k⁻¹)
                ifx ← !dealt
                tag x← dealt
                q_g.infix (x)
                else
conclude duplicate x
return UNSUCCESSFUL
```

Once n_i is updated into the list of the queue, it switches to a newer state called awaiting state.

All the n_i which are already processed are not considered again, since these are marked as tagged nodes. The mechanism for BFS-based bidirectional routing strategy is illustrated in algorithm I, and the diagrammatical illustration has been showcased in Figure 4.8.

4.6 Opportunistic Neuro-Fuzzy System

Recent advancements in critical WSN applications based on the neuro-fuzzy system extended for a new research domain for a long run to improvise the autonomous process in-terms of accuracy. The idea of employing both learning and refined decision-making procedures is found to be a perfect alternative to automate any modern critical system.

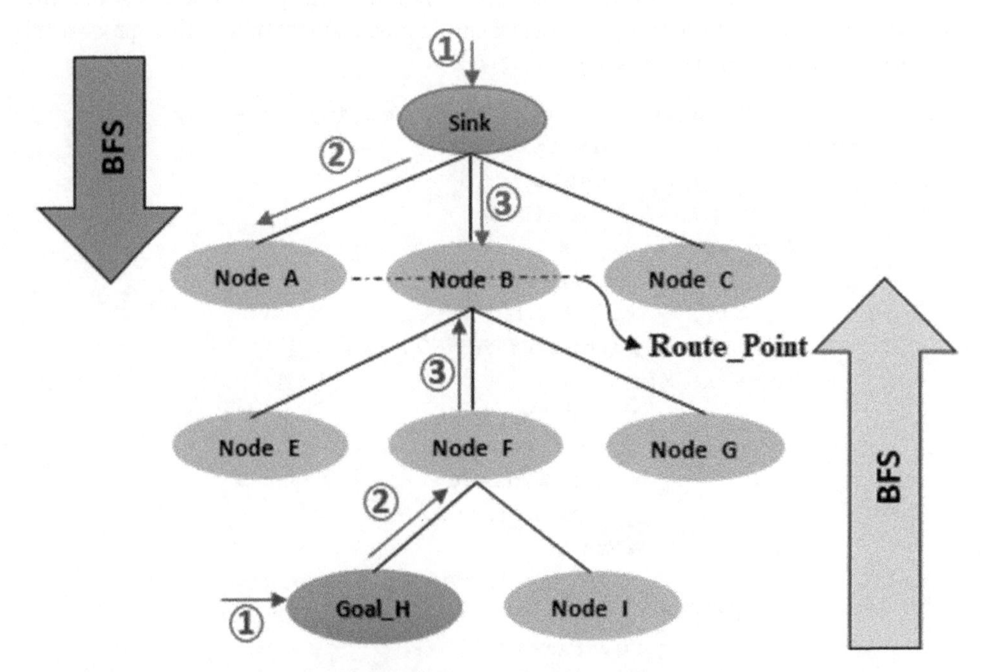

Figure 4.8 Example scenario for depicting BFS-based bidirectional routing strategy.

Thus, the demerits of the fuzzy system and neural network are evened off by combining the meritable capabilities of the neuro-fuzzy system. Figure 4.9 in Section 4.6.1 represents the neuro-fuzzy architecture.

4.6.1 Neuro-Fuzzy Architecture

4.6.2 Layer 1: Input and Membership Function

Lemma 1: The assistance of fuzzy set D (referred to as $sp \rightarrow {}_D$) is the set of the degree in X where the membership function $\mu_D(x)$ is convinced (positive):

$$sp_{\rightarrow D} = \{x \in X; \mu_D(x) > 1\} \tag{4.1}$$

Lemma 2: The core of a fuzzy set D specified in the universe of disclosure X, referred to as Core(D), sometimes as nucleus or kernel, which is the set of the degree in X where the membership function $\mu_D(x) = 1$, that is:

$$\text{Core}(D) = \{x \in X; \mu_D(x) = 1\} \tag{4.2}$$

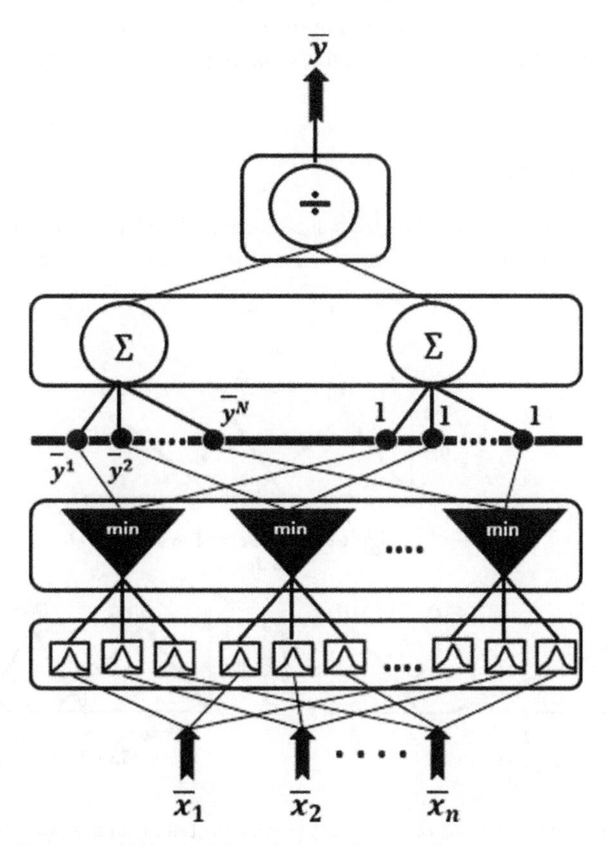

Figure 4.9 The architecture of Mamdani-based neuro-fuzzy system.

In the proposed model, a triangular membership function is considered for the neuro-fuzzy system, and it can be defined for any fuzzy set as:

$$\mu_{D^k}(x) = \begin{cases} 1 - 2\left| x - \bar{x}^k \right| / w^k & \text{if } \left| x - \bar{x}^k \right| \leq w^k / 2 \\ 0 & \text{, otherwise} \end{cases} \tag{4.3}$$

Here, $\bar{x}^k$ and w^k indicate the parameters of both the center and width of the membership function, respectively, which has been defined in Equation (4.3). According to the definition in *Lemma* 1, the "center" is assumed to be a single-point core of the triangular membership functions, whereas according to *Lemma* 2, the "width" is decided by the support (*sp*) of the membership function. Layer 1 comprises the components which substantiate the membership functions $\mu_{D_i^k}(\bar{x}_i)$, for $i = 1,\ldots,n$, and $k = 1,\ldots,N$. The crisp input values $\bar{x}_k$,.....,$\bar{x}_n$ are composed of the input vector $\bar{X} = \left[\bar{x}_k,\ldots,\bar{x}_n\right]^T$ [19]. The antecedent part of the fuzzy sets $D_1^k,\ldots,D_n^k$ for $k = 1,\ldots,N$ at the crisp degree $\mu_{D_i^k}(\bar{x}_i)$ exhibits the membership values. The count of these components is equal to *n.N*, where *n* represents the input and *N* denotes set of (IF–THEN) fuzzy rules. Thus, Layer 1 shows the membership function in the antecedent portion of the fuzzy rule. In the NFBBRR model, all the defined fuzzy sets are qualified using the triangular membership function. Since the triangular membership function expeditiously fuzzifies the crisp input in uncertain scenarios, it has been considered for inclusion in the proposed work. Figure 4.10a–c represents the triangular membership function for the three linguistic variables, namely, current energy level, average NST, and life expectancy, respectively.

4.6.3 Layer 2: Rule Evaluation

Layer 2 comprises the components which substantiate the Cartesian product (realizes the *min* operation) of the membership functional values $\mu_{D_i^k}(\bar{x}_i)$, for $i = 1,\ldots,n$. The set of *N*

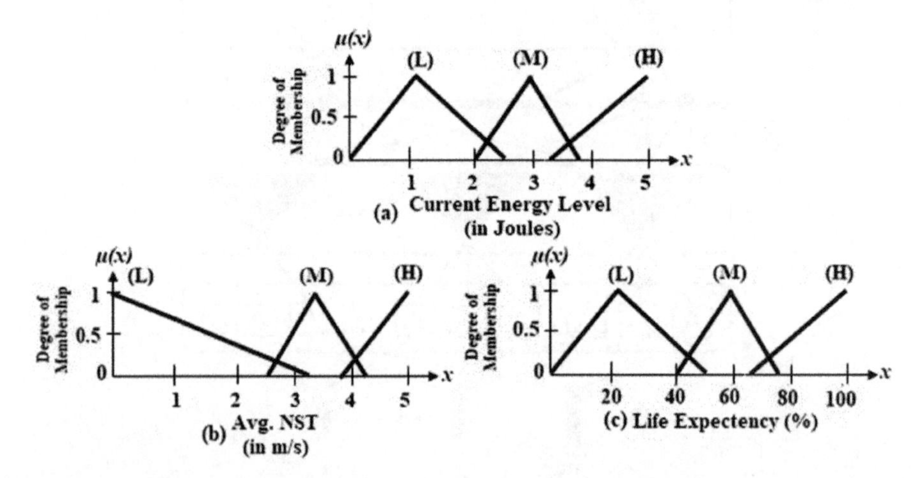

Figure 4.10 Membership functions of (a) current energy level of ni, (b) average NST, and (c) life expectancy.

elements which are composed in this layer corresponds to a particular framed rule . The set of rules in the rule base is expressed in the equation as:

$$r^{(k)}:\underbrace{\mathbf{IF}\,x_1\,\text{is}\,D_1^k\ \text{AND}\,x_2\,\text{is}\,D_2^k\ \text{AND}\ ....\ \text{AND}\,x_n\ \text{is}\ D_n^k}_{\textbf{Antecedent}}\ \mathbf{THEN}\ \underbrace{y\,\text{is}\,F^k}_{\textbf{Consequent}} \tag{4.4}$$

where $k = 1,...,N$ and $\overline{x}_k,....,\overline{x}_n$ and y are referred to as *linguistic variables* of their correspondent input and output, respectively. Furthermore, these variables are intended to fetch their values from the predefined sets that are based on different linguistic terms. In Equation (4.4), the linguistic terms are denoted as $D_1^k,....,D_n^k$ and F^k.

4.6.4 Layers 3 and 4: Defuzzification and Crisp Output Estimation

From Figure 4.7, the last two layers specify the defuzzification layers. For defuzzification, mean of maximum is employed, which produces the quantifiable output based on the fuzzy sets, $\overline{D}^k$, $k = 1,...,N$. The equation defines the process of defuzzification:

$$x^* = \sum x_i \in C^{x_i} / |C| \tag{4.5}$$

Here, C equalizes the height of the fuzzy set D, and $|C|$ represents the cardinality of the set D. Moreover, from Figure 4.7, the degree of antecedent matching is referred using retroflex hook notation and is defined as $\tau_k = \mu_{\overline{D}}k\left(\overline{y}^k\right)$. Then again, this τ_k is the primary cause of the process in the defuzzification layer because it is the output of the second layer, and the same has been given as input to the defuzzification layer. The values of $\overline{y}^k$ are propagating from the first layer and play a vital role as weights in the neural system. The component of the last layer performs the division operation to render a crisp output, $\overline{y}$, and the process is expressed as:

$$\overline{y} = \sum_{k=1}^{N} y^{-k}.\tau_k / \sum_{k=1}^{N} \tau_k \tag{4.6}$$

4.7 Performance Evaluation

4.7.1 Estimation of the Mobility of ni

Most frequently, random waypoint mobility model [20–22] is used as a mobility model in dynamic WSNs. The deployed n_i moves randomly to the opted destination with randomized velocity. Moreover, it is to be noted that the motion of n_i is independent (in terms of destination, direction, and speed) to all other deployed nodes in the network. Because of the simplicity of this model and its widespread usage, it has been considered as the preferred mobility model for the execution of the proposed system. Later on, Johansson and Larsson [23] framed the mobility metric through which the average node stability time (pause time) and speed are approximated. Foremost, on utilizing the velocity (V), the relative speed (rs) between a pair of nodes i and j at a given time t is equated as:

$$rs(i,j,t) = \left|\vec{V}i(t) - \vec{V}j(t)\right| \tag{4.7}$$

Now, the mobility metric M is equated [20] as:

$$M = \frac{1}{|i,j|} \sum_{i=1}^{N} \sum_{j=j+1}^{N} \frac{1}{\text{ST}} \int_{0}^{\text{ST}} \text{rs}(i,j,t)\,\text{d}t \tag{4.8}$$

where $|i,j|$ denotes the number of distinct node pair (i,j), N is the total number of sensor nodes deployed in the network, and ST represents the total simulation time. The average NST of n_i is computed as:

$$\text{Avg.NST}(n_i) = \sum_{i=1}^{n} \frac{M_{n_i}}{N} \tag{4.9}$$

4.7.2 Estimation of Life Expectancy of *ni*

The involvement of n_i for various operational activities (receiving data, transmitting data, listen for data, and node motion) gradually shrinks its LE. From Reference [24], it was noted that the LE of any n_i is dependent on its involvement in active communication. Such involvement will decrease the LE of a particular n_i at a varying time t. At the initial deployment, the LE of n_i is assumed to be 100%. Based on all these facts, the LE of n_i is computed as:

$$\text{LE}_{n_i}(t) = 100 - \left[\sum_{t=0}^{(\text{ST}-\Delta t)} \frac{\left(\text{TTD}_{T_x} + \text{TTD}_{R_x} + \text{TPT}\right)}{\left(\text{ST} - \Delta t\right)} \right] \tag{4.10}$$

where TTD_{T_x} and TTD_{R_x} indicate the total time duration of n_i involved in active communication (transmission and reception), and TPT_{n_i} at a given time t represents the total pause time of n_i, which is computed from Equation (4.11):

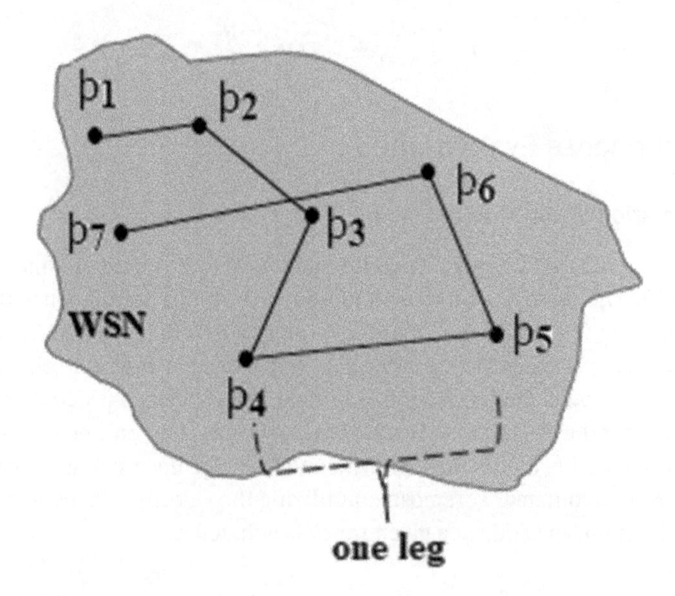

Figure 4.11 Rule inference.

$$\mathrm{TPT}_{n_i}(t) = \sum_{t=0}^{(\mathrm{ST}-\Delta t)} \left[\frac{\text{þ}_i}{(\mathrm{ST}-\Delta t)} \right] \quad (4.11)$$

The *pause time* is defined as the waiting period of the node n_i at each waypoint before moving to the next þ$_i$, sometimes referred to as one *leg*.

4.7.3 Representation of Rule Base

The rule inference is performed based on the Mamdani inference [25] method, which follows simple "IF 'A' is X AND 'B' is Y, THEN 'C' is Z" approach [16]. As a result, all three input variables together tend to form 27 rules in the rule base to determine the n_i possibilities for routing the essential data. For better understanding the purpose, Figure 4.11 demonstrates the rule base and their possible rule firing. Finally, the framed rules are categorized for decision-making and training purposes during route establishments. The corresponding n_i that executes neuro-fuzzy inferring methods tends to identify the optimal route node. In case of the least possibilities of obtaining an optimal route node, as a contingency plan, the second variant of the favorable node-set is opted for the routing purpose. Thus, such a type of contingency plan indirectly includes the fault-tolerance capabilities of the network, especially during the event of data transmission. The rule categorization for determining best and favorable nodes is constituted in Table 4.2.

4.7.4 QoS Comparison and Analysis

To constitute the potency of NFBBRR, performance analysis has been performed using the NS-2 simulation tool [26]. The primary necessity of simulation is to execute NFBBRR to experiment, evaluate, and compare the evaluation reports of the nominated approach with well-known recent and advanced approaches.

For simulation, a range of 20–100 nodes is considered in the network area of 1500 × 1500 m^2, and for various experimental scenarios, the total simulation time is fixed to be 600 seconds. Since the node is mobile, they are configured with a speed of 0–10 m/s, and the random waypoint model is employed as a mobility model. Both the transmission range and initial energy are configured to be 250 meters and 300 joules, respectively. User Datagram Protocol (UDP)–based constant bit rate is chosen to be the traffic type.

To assess the performance of the NFBBRR approach, it is compared with the most recent and advanced routing protocols such as fuzzy TOPSIS rough set analytical hierarchical

Table 4.2 Rule assortment and outcomes

Rule categorization	Outcome
R13, R14, R16, R17, R22, R23, R25, R26	Highly favored routing nodes (best)
R10, R11, R19, R20	Favorable routing nodes
All remaining rules	Riskiest routing nodes

process (FTR-AHP) [27], energy-efficient routing approach (EE-RA) [28], and genetic algorithm (GA) [29]. For precise evaluation, some of the most prominent and standard metrics like throughput, packet delivery ratio (PDR), end-to-end delay, network overhead, and average energy consumptions [30] are considered against the varying network size and varying pause time of the deployed node.

Throughput is usually defined as the number of bytes received successfully per single packet transmission time, and it is quantified in kbps. It is observed that the throughput of each approach is steadily increasing as the network size increases. Among the four approaches, NFBBRR shows a prominent impact due to its fast route establishment strategy. The network gains convenient throughput when the deployed nodes have high pause time in their randomized movement from one point to another. This effect is due to the intermediate stability of the node's motion and multihop transmission options, as the scalability of the network increases. In both cases, the proposed approach shows significant improvements to all other existing approaches.

From the previous studies, it is found that whenever there is a decrease in the node's pause time and an increase in network size, both NFBBRR and ANFIS-based EE-RA deliver better results as compared to other existing approaches.

It is already examined that, in EE-RA, on account of optimal hello message fraction, the hello interval has been significantly reduced with increase in network size and decrease in pause time. But still, EE-RA lacks in gaining essential controlling effects like NFBBRR. Thus, under various conditions, the proposed system controls the node effectively than the existing methods.

The average EC of the network is calculated as:

$$\text{Avg. EC}\left(\text{in joules}\right)=\frac{\text{Total energy consumption}}{\text{Total number of deployed nodes}} \tag{4.12}$$

It is observed that the EC increases as the speed of the node mobility increases in all the approaches. This is because the node has to spend additional energy to update its new neighbors to the current speed of each mobile node. In both cases, the proposed NFBBRR model comes out as the least energy consuming model, since it utilizes both the source and sink nodes to construct the routing path. Further, the origin of path establishment from both the ends (source and sink) quickly accomplishes the task.

End-to-end delay concerns the total time spent by the packets to arrive at the destination from the source across a network. There are some prominent impactful parameters such as queuing, packet processing, and transmission, almost all of which impact the network delay. It has been observed that, whenever the scalability of the network increases with randomized nodes mobility, the performance of the NFBBRR approach gives more beneficial results as compared to EE-RA, FTR-AHP, or GA. This effect is due to the simultaneously routing action of both source and sink nodes in a routing path. For most of the generated events, this metric tends to qualify the correctness and completeness of the routing approach. Thus, NFBBRR renders the path optimality for each generated event (either normal or critical).

PDR is important, as it draws the loss rate especially in the transport layer, which in turn leads to the minimization of throughput gain of the entire network. Similar to end-to-end delay, this metric also stipulates the completeness of any routing protocol. If the transmission packets are routed in the utmost minimum time, then the delivery and receiving ratio of the data packets within a simulative limited timespan may eventually increase. Thus, the PDR of the network gradually increases as the size of the network scales in an incremental mode, whereas the resultant of PDR at the higher pause time of a mobile node implies that the stable point of mobile nodes has increased the delivery of data packets without much packet drops. Based on the overall comparison of each approach, NFBBRR has again exhibited outstanding performance than the existing approaches.

The primary goal here is to attain fast data deliverance in a limited time span, where the average deliverance time span of data transmission from source to sink among NFBBRR, FTR-AHP, EE-RA, and GA on the occurrence of an event (either critical or normal) is compared. From the result, we can note that the proposed NFBBRR model outperforms the other models, since the routing path construction happens simultaneously at both the ends of event generation and event requisition. The average time span of NFBBRR required to complete the transmission is around 1.99 seconds, whereas it is around 2.17 seconds for other models, which shows a percentage difference as 8.65%. This difference can make major impacts, especially when concerning critical environments.

4.8 Conclusion

This chapter demonstrates a fast route establishment and energy-efficient routing approach for handling the critical events in a dynamic network. The quick route establishment enables fast data deliverance and optimizes the frequent needs of the required information at the sink end. This is done by employing a neuro-fuzzy-based bidirection, biobjective reactive routing protocol. One of the biobjectives, namely, fast route establishment in the dynamic network, is characterized by the determination of the optimal routing node, which is achieved by utilizing the primary input parameters (current energy level and life expectancy) in the neuro-fuzzy inferring method. Further, the impartation of BSF-based bidirectional path construction enhances network stability under various dynamic conditions of the mobile nodes. Simulation observation records and the examination of the NFBBRR model in comparison with the other existing models showcase its outstanding performance. Thus, as demonstrated in Tables 4.3 and 4.4, the proposed approach accomplishes better resultants in terms of E-to-E delay, NO, avg. EC, PDR, and throughput on applying a variation of pause time and network size. Maintaining fast data deliverance across the network against variations of node mobility is a challenging process that has been considered as a major problem and is resolved through the proposed strategy.

The average complete event deliverance time of NFBBRR outsmarts the other models by employing a bidirectional routing strategy. Besides these, on the comparative results, the proposed model exhibits 6.92% and 15.46% of utilization reduction in energy consumption against varying network size and pause time, respectively. Similarly, it demonstrates a reduction of 8.65% and 39.08% in network overheads against varying network size and pause time, respectively. In the case of PDR and throughput, the proposed model shows an increase of 5.48% and 23.52%, respectively, as compared to other models.

Eventually, in future work, this research work can be employed in a distributed dynamic smart network, where nodes may opt for a multihop routing strategy with limited resource restriction.

Table 4.3 Comparative analysis on the influential effect of network scalability.

Metrics / Network size	25				50				75				100			
NO	0.4	0.5	0.6	0.4	0.71	0.75	0.74	0.72	0.65	0.78	0.8	0.84	0.72	0.8	0.88	0.9
Avg. Ec (in J)	15.3	16.2	17.1	16.7	15.9	16.4	17.3	16.9	16	16.6	17.6	17.4	16.4	16.5	18.2	17.6
E-to-E delay (ms)	187	295	264	341	150	152	164	167	135	120	125	130	175	350	400	435
PDR (%)	80.56	78.14	74.6	70.4	84	81.23	80	77.23	86.23	82.44	83.12	81	94.23	90.12	89.1	92.12
Throughput (kbps)	147	100	125	133	179	124	136	148	289	212	198	208	345	311	301	278
Protocols	NFBBRR	FTR-AHP	EE-RA	GA	NFBBRR	FTR-AHP	EE-RA	GA	NFBBBRR	FTR-AHP	EE_RA	GA	NFBBRR	FTR-AHP	EE-RA	GA

Table 4.4 Comparative analysis on the influential effect of pause time of mobile nodes.

Metrics \ Pause time	300				240				180				120				60			
NO	0.19	0.24	0.38	0.4	0.2	0.3	0.4	0.5	0.24	0.35	0.48	0.6	0.22	0.4	0.5	0.67	0.19	0.24	0.38	0.4
Avg. Ec (in J)	2.1	2.3	2.2	2.4	3.1	3.2	3.4	3.7	3.4	4.2	4.9	4.3	4.5	5.2	5	5.1	4.8	5.7	5.8	5.4
E-to-E delay (m^3)	15.4	16.23	17.2	18.4	16.4	14.2	15.64	14.87	11.2	12.45	13.54	13.12	9.4	10.23	10.1	10.9	7.4	9.1	8.45	8.1
PDR (%)	98.1	91.1	88.2	90.1	99.1	92	90.1	92.2	94.3	88.2	86.4	88.74	95.2	87.3	84.1	89.2	92.33	82.12	82.87	84.31
Throughput (kbps)	278	258	265	287	214	201	198	187	185	164	181	178	134	128	139	143	98	94	85	80
Protocols	NFBBRR	FTR-AHP	EE-RA	GA	NFBBRR	FTR-AHP	EE-RA	GA	NFBBBRR	FTR-AHP	EE_RA	GA	NFBBRR	FTR-AHP	EE-RA	GA	NFBBRR	FTR-AHP	EE-RA	GA

Data Availability

The data utilized in supporting the examination and observations of this study are available from the corresponding author upon request.

Conflicts of Interest

There are no conflicts of interest to declare in this chapter.

References

[1] I.F. Akyildiz, W. Su, Y. Sankarasubramaniam, and E. Cayirci, "Wireless sensor networks: A survey," *Comp. Netw.*, vol. 38, no. 4, pp. 393–422, 2002.

[2] C. Gómez-Calzado, A. Casteigts, A. Lafuente, and M.A. Larrea, "Connectivity model for agreement in dynamic systems," In *21st International Conference on Parallel and Distributed Computing*, Vienna, Austria, 2015; pp. 24–28. doi:10.1007/978-3-662-48096-0_26.

[3] C. Gómez-Calzado, "Contributions on agreement in dynamic distributed systems," Ph.D. Thesis, Universidaddel País Vasco-Euskal Herriko Unibertsitatea, Leioa, Vizcaya, Spain, 2015.

[4] T.A. Muthupandian, J.G. Eanoch, and H. Robinson Yesudhas, "A survey on techniques for selection of forwarding node in wireless sensor networks," *Int. J. Adv. Comp. Electr. Eng.*, vol. 2, no. 4, pp. 24–29, 2017.

[5] R. Pon Rohini, S. Shirly, and D.C. Joy Winnie Wise, "Multipath routing using neuro-fuzzy in wireless sensor network," *Int. J. Res. Appl. Sci. Eng. Tech.*, vol. 3, no. 4, pp. 331–333, 2015.

[6] K. Sasikala and V. Rajamani, "A neuro-fuzzy based conditional shortest path routing protocol for wireless mesh network," *Int. J. En. Res. Manag. Comp. Appl.*, vol. 2, no. 5, pp. 1–10, 2013.

[7] S.R. Chandran, V.S. Manju, and A.P. Alex, "A neuro-fuzzy approach to route choice modelling," *Int. J. Sci. Appl. Inf. Tech.*, vol. 2, no. 2, pp. 9–11, 2013.

[8] E. Golden Julie, S. Tamil Selvi, and Y. Harold Robinson, "Performance analysis of energy efficient virtual back bone path-based cluster routing protocol for WSN," *Wireless Pers. Comm.*, vol. 91, no. 3, pp. 1171–1189, 2016, Springer.

[9] S. Basagni, A. Carosi, E. Melachrinoudis, C. Petrioli, and Z.M. Wang, "Controlled sink mobility for prolonging wireless sensor networks lifetime," *Wireless Netw.*, vol. 14, no. 6, pp. 831–858, 2008.

[10] I. Chatzigiannakis, A. Kinalis, and S. Nikoletseas, "Sink mobility protocols for data collection in wireless sensor networks," In *Fourth ACM International Workshop on Mobility Management and Wireless Access*, pp. 52–59, 2006. doi:10.1145/1164783.1164793.

[11] A.T. Erman, L. Van Hoesel, P. Havinga, and J. Wu, "Enabling mobility in heterogeneous wireless sensor networks cooperating with UAVs for mission-critical management," *IEEE Wireless Comm.*, vol. 15, no. 6, pp. 38–46, 2008.

[12] S. Deng, J. Li, and L. Shen, "Mobility-based clustering protocol for wireless sensor networks with mobile nodes," *IET Wireless Sens. Sys.*, vol. 1, no. 1, pp. 39–47, 2011.

[13] P. Suriyachai, U. Roedig, and A. Scott, "A survey of MAC protocols for mission-critical applications in wireless sensor networks," *IEEE Comm. Surv. Tutor.*, vol. 14, no. 2, pp. 240–264, 2012.

[14] W. Ye, J. Heidemann, and D. Estrin, "An energy-efficient MAC protocol for wireless sensor networks," *21st Annual Joint Conference of IEEE Computer and Communications Societies*, New York, USA, 2002; vol. 3, pp. 1567–1576. doi:10.1109/INFCOM.2002.1019408.

[15] M. Elleuch, K. Heni, and A. Mohamed, "Exploiting neuro-fuzzy system for mobility prediction in wireless ad-hoc networks," In *International Work-Conference on Artificial Neural Networks, Palma de Mallorca*, Springer, Spain, 2015; pp. 536–548. doi:10.12178/1001-0548.2019076.

[16] K.M. Karthick Raghunath and S. Thirukumaran, "Fuzzy-based fault-tolerant and instant synchronization routing technique in wireless sensor network for rapid transit system," *Automatika: J. Cont. Meas., Electr. Comp. Comm.*, vol. 60, no. 5, pp. 547–554, 2019.

[17] K. Nikhil, S. Agarwal, and P. Sharma, "Application of genetic algorithm in designing a security model for mobile adhoc network," *Int. Conf. Inf. Technol. Converg. Serv.*, vol. 2, pp. 181–187, 2012. doi:10.5121/csit.2012.2116.

[18] H. Kumar, S.M.D. Kumar, and G.P. Sunitha, "Optimal multipath routing using BFS for wireless sensor networks," *J. Netw. Comm. Emerg. Tech.*, vol. 7, no. 3, pp. 22–29, 2017.

[19] D. Rutkowska, *Neuro-Fuzzy Architectures and Hybrid Learning*, Springer-Verlag, New York, 2002.

[20] F. Bai, N. Sadagopan, and A. Helmy, "The important framework for analyzing the impact of mobility on performance of routing for ad hoc networks," *AdHoc Netw. J. Elsevier Sci.*, vol. 1, no. 4, pp. 383–403, 2003.

[21] J. Broch, D.A. Maltz, D.B. Johnson, Y.-C. Hu, and J. Jetcheva, "A performance comparison of multi-hop wireless ad hoc network routing protocols," *ACM/IEEE international conference on Mobile computing and networking*, 1998; pp. 85–97, doi:10.1145/288235.288256.

[22] D.B. Johnson and D.A. Maltz, "Dynamic source routing in Ad Hoc wireless networks," In T. Imielinski and H.F. Korth (eds) *Mobile Computing, the Kluwer International Series in Engineering and Computer Science* (vol. 353, pp. 153–181), Springer, Boston, MA, 1996. doi:10.1007/978-0-585-29603-6_5.

[23] P. Johansson and T. Larsson, "Scenario-based performance analysis of routing protocols for mobile Ad-Hoc networks," In *Fifth Annual ACM/IEEE International Conference on Mobile Computing and Networking*, 1999; pp. 195–206. doi:10.1145/313451.313535.

[24] K.M. Karthick Raghunath and N. Rengarajan, "Response time optimization with enhanced fault-tolerant wireless sensor network design for on-board rapid transit applications," *Clus. Comp. J. Netw. Softw Tools Appl.*, vol. 22, no. July-S4, (2019), pp. 9737–9753, 2017.

[25] L.P. Perera, J.P. Carvalho, and C.G. Soares, "Solutions to the failures and limitations of Mamdani fuzzy inference in ship navigation," *IEEE Trans. Veh. Tech.*, vol. 63, no. 4, pp. 1539–1554, 2014.

[26] "The Network Simulator ns-2," Information Sciences Institute, USA. Viterbi School of Engineering, September 2004. Available on: http://www.isi.eu/nsnam/ns

[27] S.R.M. Krishna, V. Kamakshi Prasad, and M.N. Seeta Ramanath, "Optimal reliable routing path identification in MANET with FTR-AHP model," In *IEEE Systems Conference (SysCon)*, Orlando, FL, USA, 2016. doi:10.1109/SYSCON.2016.7490633.

[28] D. Bisen and S. Sharma, "An energy-efficient routing approach for performance enhancement of MANET through adaptive neuro-fuzzy inference system," *Int. J. Fuzzy Sys.*, vol. 20, no. 8, pp. 2693–2708, 2018.

[29] A. Roy and S.K. Das, "QM^{2}RP: A QoS-based mobile multicast routing protocol using multi-objective genetic algorithm," *Wireless Netw.*, vol. 10, no. 3, pp. 271–286, 2004.

[30] D. Bisen and S. Sharma, "An enhanced performance through agent-based secure approach for mobile ad-hoc networks," *Int. J. Electr.*, vol. 105, no. 1, pp. 116–136, 2018.

5

Feature Detection and Extraction Techniques for Real-Time Student Monitoring in Sensor Data Environments

Dr. V. Saravanan and Dr (Ms). N. Shanmuga Priya

Department of Computer Applications (PG), Dr. SNS Rajalakshmi College of Arts and Science, Coimbatore, India

5.1 Introduction

The research mainly focuses on the potential of student motion behavior analysis. This study is conducted for the learning of repeated motion behavior with respect to the students (i.e., the frequently visited places and the paths taken between the places) and thereafter to show that it is possible to detect unusual behavior using the knowledge of frequent behavior [1]. The best example for this scenario is taking a wrong route and getting lost. An important objective of pervasive computing is to give accurate information about people behavior. It has a wide range of applications such as in medicine, security solutions, and student monitoring in educational campuses [2].

Computer vision is a technology in which complex and tiny sensors such as video cameras are used to capture human motion [3]. The detection of human motion is a promising area of research. Sensors that are embedded in wearable objects are attached to the body, which is studied for its patterns or behavior. The data generated are normally from GPS sensors. The need of the hour is to process these data using signal-processing methods and to recognize patterns of real-time human motion [4]. This chapter presents a novel framework using deep support vector machines (SVMs) for the detection and extraction of features from real-time sensor data obtained through the wireless sensor networks (WSNs) placed inside an academic campus. The movements of the students are then captured and stored using a small peripheral device. The following sections of this chapter are as follows: Section 5.2 discusses the existing techniques used for human motion detection; Section 5.3 presents the proposed methodology; and Sections 5.4 and 5.5 discuss the experimental setup and the results obtained. The chapter concludes with Section 5.6.

5.2 Existing Works

A few human movement acknowledgment frameworks have been proposed previously, which incorporate the utilization of accelerometers. Some of them examine and group various types of actions utilizing increasing speed signals [5], while focus on perceiving a wide

Sensor Data Analysis and Management: The Role of Deep Learning, First Edition. Edited by A. Suresh, R. Udendhran, and M.S. Irfan Ahmed.
© 2021 John Wiley & Sons, Ltd. Published 2021 by John Wiley & Sons, Ltd.

arrangement of day-to-day physical exercises, or portray human activity recognition (HAR) in view of highlight determination strategies [6, 7]. Bernecker et al. [7] proposed a renaming identification step that builds exactness of movement acknowledgment. Karantonis et al. [8] presented an on-board handling method for an ongoing grouping framework, yielding outcomes that exhibited the possibility of executing an accelerometer-based, constant development classifier utilizing implanted knowledge. Khan et al. [9] proposed a framework that utilizes a various-leveled acknowledgment conspire, that is, the state acknowledgment at the lower level utilizing factual highlights and the movement acknowledgment at the upper level utilizing the increased element vector followed by direct discriminant investigation [10]. A few incredible calculations have been proposed in the studies for human movement recognition. The most generally utilized are counterfeit neural systems [11], the credulous Bayes [12], and the help vector machines [13].

5.3 Proposed Methodology

The proposed methodology consists of foreground and background detection, which is done using improved algorithms. The second step is the process of feature selection using deep code generation convolutional neural network (CNN), which is followed by outlier detection using SVM-based classification.

5.3.1 Foreground and Background Detection

The improved method OTSU, which is basically a classical image segmentation technique in which clustering principle is utilized, is one of the appealing solutions for the stability problem. The probability of misclassification is alleviated by classifying the gray values in between the parts to a maximum based on the calculated variance in order to find an optimum gray level. The improved ViBe algorithm updates the sample by estimating the valid sample value probability, and the sample value of the background is constructed henceforth. With respect to these characteristics, the first sample value is chosen as the foundation outline at time "*t*." At that point, the difference in between the foundation and the present scenes is calculated and represented as $D(x, y)$. The differential motion picture of $D(x, y)$ has the closest view and the foundation that are apt for OTSU. The degree of $d(x, y)$ is set as L. The pixels are then arranged in a couple of classes, namely, $\{0, 1, T\}$ $and\{T, T+1, \ldots, 255\}$. The calculation of variance in between the two categories is calculated as:

$$\sigma^2 = P_0\left[\mu_{0-}u\right]x^2 + P_1\left[\mu_{1-}u\right]y^2. \tag{5.1}$$

P_0 and P_1 speak to the two sorts of pixels, individually. Variable u speaks to the normal dark, and u_0 and u_1 speak to the mean gray estimation of the two sorts of pixels, separately. The more noteworthy the estimation of pixel (5) is, the better the edge of the division. Utilizing equation (6), the ideal estimation of the between-class difference of t^* is the ideal powerful limit after the entire picture is navigated. The principal edge of the choosing test is utilized as the foundation outline, so as to maintain a strategic distance from excessively huge or too little edge; the predisposition edge of t^* is constrained. T_{Min} and T_{Max} denote the upper and lower edges, respectively:

$$t^* = Arg\ \ max\ \ \sigma^2. \tag{5.2}$$

The postprocessing is improved in the following two aspects once the moving object is distinguished by the improved ViBe method.

5.3.2 Deep Binary Code Generation CNN

Deep learning (DL) is a kind of advanced artificial neural network (ANN) and was introduced for machine learning (ML) methods to address a dissimilar degree of issues in classification problems. The main job of the DL procedure is to blackmail the data in more DL.

For several indistinguishable motions, a high reaction includes maps seen at similar record areas in the profound layer. Based on the recognition, the portrayal of the picture is then transmuted as a paired code. This code is created through differentiation of the reaction that starts at every component map with that of the normal reaction on all the elementary maps:

$$a_k^{\text{HL}} = \sigma(a_k^7 W^{\text{HL}} + b^{\text{HL}}). \tag{5.3}$$

Here, $\sigma(a_k^7 W^{\text{HL}} + b^{\text{HL}})$ represents the sigmoid function that regulates the output in between the interval [0,1], and a_k^7 represents the output features' vectors in the Fc7 layer. W^{HL} represents the weight, and b^{HL} represents the bias of the hidden layer. This is represented as shown in Equation (5.4):

$$b_k = \begin{cases} 1, & a_k^{\text{HL}}\ \phi 0.5 \\ 0, & a_k^{\text{HL}} \leq 0.5 \end{cases} \tag{5.4}$$

5.3.3 Deep SVM for Outlier Detection

Deep SVM classification is one of the best solutions for an efficient vehicle-tracking system once the effective moving masses are extracted. Also, it has the advantage of eliminating the interference of nonmotor vehicles or pedestrians. The superior property of SVM-based classifiers on nonlinear and large dimensional space aids in identifying the tracking regions effectively. The rotation invariance and gray invariance properties of LBP add efficiency weightage to the vehicle-detection process. This process involves the following steps:

- First, video samples and negative samples are gathered, and the LBP features are identified and extracted.
- The SVM is performed for the classification.
- DL is applied to identify the outliers.
- The regions that fall in the learned region are left out, outliers are considered as abnormal behavior, and alerts are set accordingly.

5.4 Experimental Setup

Privacy is one of the major concerns when a subject is being monitored. Hence, the domain of analysis is kept local on the subject and restricted to a small peripheral device. The experiments are carried out on the NS2 simulator tool. The number of nodes (students) is taken as 100 with a domain area of 100 m^2. The network life time is chosen as 15 seconds.

5.5 Results and Discussion

Table 5.1 shows the comparative analysis of the proposed method against that of the other methods in terms of performance (precision, recall, and F-measure).

Figure 5.1 shows the pictorial representation of Table 5.1, and it is evident that the proposed deep SVM method outperforms the other methods discussed in the literature. Table 5.2 shows the comparative analysis of the proposed method in terms of accuracy measures—the mean relative error (MRE) and mean absolute relative error (MARE).

Figure 5.2 shows the graphical representation of Table 5.2. It is evident from the result that the proposed deep SVM method has very low MRE and MARE values, which ensures more accuracy than the other methods discussed in the literature.

Table 5.1 Comparative analysis of the proposed method.

Method	Performance measures		
	Precision	Recall	F-measure
HARD	0.87	0.89	0.85
RCD	0.83	0.82	0.85
OBPD	0.90	0.92	0.94
HRS	0.67	0.69	0.71
LDA	0.76	0.78	0.79
Deep SVM	0.93	0.94	0.96

Table 5.2 Comparison of accuracy.

Method	Accuracy measures	
	MRE	MARE
HARD	0.41	0.43
RCD	0.19	0.21
OBPD	0.31	0.31
HRS	0.29	0.29
LDA	0.21	0.22
Deep SVM	0.12	0.16

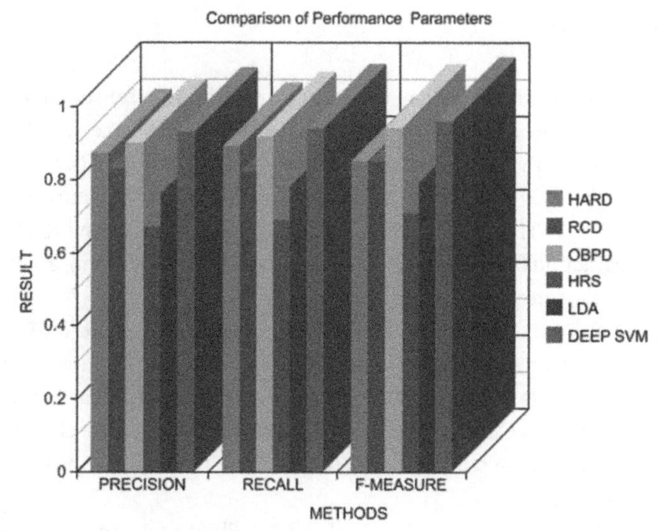

Figure 5.1 Comparison of performance parameters.

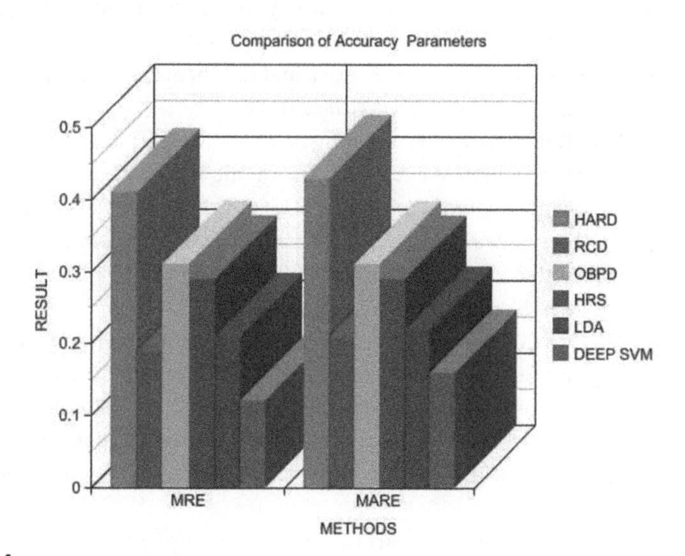

Figure 5.2 Comparison of accuracy parameters.

5.6 Conclusion

A novel technique for feature detection and extraction based on deep SVM is presented. This study specifically focuses on student motion behavior analysis, which is carried out using improved OTSU and deep binary code generation methods. The deep Support Vector Machine method is proposed for the classification of normal behavior and outliers. The experiments were carried out using GPS sensor data in the NS2 simulator. The results obtained through the experiments prove that the proposed technique outperforms the other methods in terms of performance and accuracy.

References

[1] J. Hunter and M. Colley, "Feature extraction from sensor data streams for real-time human behaviour recognition," In J.N. Kok, J. Koronacki, R. Lopez de Mantaras, S. Matwin, D. Mladenic and A. Skowron (eds) *Knowledge Discovery in Databases: PKDD 2007* (vol. 4702). Springer, Berlin, Heidelberg, 2007. Lecture Notes in Computer Science.

[2] J. Yan, F. Tian, H. Qinghua, Y. Shen, X. Shan, J. Feng, and K. Chaibou, "Feature extraction from sensor data for detection of wound pathogen based on electronic nose," *Sens. Mater.*, vol. 24, no. 2, pp. 57–73, 2012.

[3] O. Politi, I. Mporas, and V. Megalooikonomou, *Comparative Evaluation of Feature Extraction Methods for Human Motion Detection*. Multidimensional Data Analysis and Knowledge Management Laboratory University of Patras 26500 Rion-Patras, Greece.

[4] L. Wang, T. Gu, X. Tao, and J. Lu, "A hierarchical approach to real-time activity recognition in body sensor networks," *J. Perv. Mob. Comp.*, vol. 8, no. 1, pp. 115–130, 2012.

[5] C.-C. Chang and C.-J. Lin, "LIBSVM: A library for support vector machines," *ACM Trans. Intel. Sys. Tech.*, vol. 2, pp. 1–27, 2011.

[6] N. Krishnan and D.J. Cook, "Activity recognition on streaming sensor data," *Perv. Mob. Comput.*, vol. 10, pp. 138–154, 2014.

[7] T. Bernecker, F. Graf, H. Kriegel, and C. Moennig, "Activity recognition on 3D accelerometer data," *Tech. Rep.*, 2012.

[8] D.M. Karantonis, M.R. Narayanan, M. Mathie, N.H. Lovell, and B.G. Celler, "Implementation of a real-time human movement classifier using a triaxial accelerometer for ambulatory monitoring," *IEEE Inf. Tech. Biomed.*, vol. 10, pp. 156–167, 2006.

[9] A. Khan, Y. Lee, S.Y. Lee, and T. Kim, "Triaxial accelerometer-based physical-activity recognition via augmented-signal features and a hierarchical recognizer," *IEEE Trans. Inf. Tech. Biomed.*, vol. 14, no. 5, pp. 1166–1172, 2010.

[10] M. Zhang and A. Sawchuk, "USC-HAD: A daily activity dataset for ubiquitous activity recognition using wearable sensors," In *UbiComp '12*, USA, 2012. Conference on Ubiquitous Computing, September 2012, pp. 1036–1043, doi: 10.1145/2370216.2370438.

[11] A.M. Khan, Y.K. Lee, and S.Y. Lee, "Accelerometer's position free human activity recognition using a hierarchical recognition model," *In IEEE Health Communication, 2010*

[12] A. Avci, S. Bosch, M. Marin-Perianu, R. Marin-Perianu, and P. Havinga, *Activity Recognition Using Inertial Sensing for Healthcare, Wellbeing and Sports Applications: A Survey*, University of Twente, The Netherlands, 2010.

[13] T. Gu, Z. Wu, X. Tao, H. Pung, and J. Lu "epSICAR: An emerging patterns based approach to sequential, interleaved and concurrent activity recognition," 2009.

6

Deep Learning Analysis of Location Sensor Data for Human-Activity Recognition

Hariprasath Manoharan[1], Ganesan Sivarajan[2], and Subramanian Srikrishna[3]

[1] *Assistant Professor, Department of Electronics and Communication Engineering, Audisankara College of Engineering and Technology, Gudur, Andhra Pradesh*
[2] *Associate Professor, Department of Electrical and Electronics Engineering, Government College of Engineering, Salem, Tamil Nadu*
[3] *Professor, Department of Electrical and Electronics Engineering, Annamalai University, Chidambaram, Tamil Nadu*

6.1 Introduction

For improving the working efficiency of sensors and for testing them under different conditions, a predictive algorithm is essential. This is possible only when deep learning methods are used, where different strategies are followed when any problem occurs on the network. If sensors are installed, then the network depends on main node for the purpose of storing and accessing the data. For sensing the information and sending it to applications, such as those for health monitoring, there should be less delay. This robust prediction of health is necessary because it can save people's lives. There are many deep learning models integrated with neural networks, where many hidden layers are included in the shape of a cascade.

This cascaded network provides the necessary machine learning techniques for the automatic detection of datasets. Most of the useful properties from datasets are based on soft target models with probability analysis. This probability analysis is used for extracting the exact values of deep models, which minimizes interpretable learning. The decision rule and tree structures are based on the performance of complex models monitored by expert clinicians. The block diagram of integrating the sensor models is shown in Figure 6.1.

6.1.1 Existing Deep Learning Techniques in Health Care Monitoring

Since the working of sensors is limited with respect to parameters such as energy and bandwidth, it should be integrated with deep learning models [1]. Deep learning models are classified as controlled and uncontrolled techniques, where, for understanding different perspectives, both the methods are used for creating green energy solutions. Also, reference [1] suggests that the applications of deep learning can be used for creating an energy awareness situation

Sensor Data Analysis and Management: The Role of Deep Learning, First Edition. Edited by A. Suresh, R. Udendhran, and M.S. Irfan Ahmed.
© 2021 John Wiley & Sons, Ltd. Published 2021 by John Wiley & Sons, Ltd.

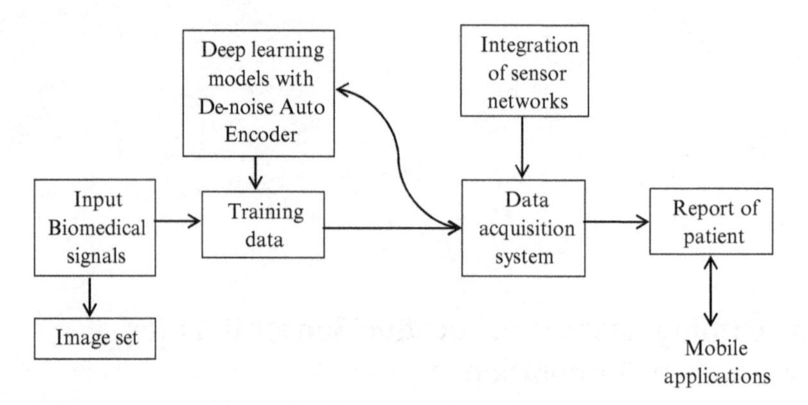

Figure 6.1 Conventional deep learning with sensor networks.

if the resources are properly allocated. For computational phenotype diseases [2], a healthy prediction methodology is necessary, which is possible only when the sensors are located in the human body. However, the sensor needs to be expanded with a knowledge technique for analyzing the effects present inside the human body, which is possible only when deep learning models are used [2]. This healthy prediction is prepared by integrating the dataset in a series of time, which produces the output for the decision-making process.

It is important to know both the volume and velocity of the sensor that has to be implemented for healthcare prediction [3], because activity recognition is based on these parameters. To know the volume and velocity, the sensors are combined with neural models where the rate of classification is improved to a much greater extent. If the classification rate is improved, then the relationship between activity and features can be clearly established [3]. Therefore, the exact mechanism inside the neural networks can be understood clearly. The voice recognition system is one of the most important parameters while identifying human activity. It is recognized that, for communication, the primary source used is the voice, and voice degradation occurs in case of any disorder [4]. Therefore, there is a possibility that the entire system will be disturbed if the energy of the system is not monitored. For voice recognition, an automatic disorder system is designed with direct prediction, and the accuracy of prediction is also higher when sensors are designed for recognizing the voice of humans [4].

Even though the data collection system is established, some physical data need to be acquired, which will be possible only when the wearable sensors are integrated with neural network deep learning process. For establishing the physical data, a supportive vector machine formulation has been implemented [5], where the analysis is performed through an online process. During this decline process, if any emergency situation occurs, then actions will be taken immediately for solving it, and if there is any error in the physical data, this will be quickly rectified. However, it is very difficult to obtain a health analysis if the environment is related to media—that is, an exact prediction of electrocardiogram signals cannot be obtained [6]. If this situation remains the same, then the process of cloud storage and some local features that are extracted by the detection method are very difficult for the doctors to recognize.

In the aforementioned case, the accuracy of the prediction method should be above 85%, since voice pathologies are used for identifying the signals. One of the major advantages of the method [6] is that two different signals are used in the process where there is a possibility to obtain a low noise ratio. It will be easy to transmit the signals to the cloud with less noise because, after transmitting, the signals need to be classified for identification purposes. Also, the feasibility of the method should be analyzed for effectively monitoring the noise ratio. If a large quantity of data is used, then there is a need for improving the prediction mechanism [7], where the Internet of things (IoT) technique has been integrated for sensing the data and for the decision-making process. Even conventional machine learning technique has been applied as a deep learning method for making the sensor representation as one of the most useful cases. But, due to certain limitations, the loss function of the sensor that is integrated with deep learning techniques has become an inconsistent feature.

For small-sized data, health can be sensed using the bottom-up approach, but the scalability measures will be limited by several rules. Due to the disadvantages of deep learning techniques, automatic classification accuracy is affected. To overcome the disadvantages, a neural network with deep forwarding technique has been implemented [7]. The working model for large data consists of a large number of inputs; therefore, it is possible that more data can be sensed accurately and in an appropriate way. Both decoding and encoding methods are used with autoencoding technique, including fully connected layers for monitoring the data using sensor techniques. For many applications, it is important to monitor the changes in human interaction, which is denoted as the interaction between a human and a robot. Classification accuracy is necessary for defining the interaction because physiological structure with environmental data needs to be specified for locating the data [8].

There is a possibility that local interactions can be merged with each other and used for representing the signal dynamics. The emotion classification becomes accurate only when the deep learning models are integrated [8]. The wireless networks are more effective when the sensors are merged with each other.

6.2 Mathematical Representation of Parameters Used in Health-Monitoring Systems

The parameters for monitoring health are carefully chosen in such a way that the primary intention is the assortment of forwarder nodes that consume less interactive expanse, minimizing the consumption of energy and path loss. The projected methodology is stated where multiple parameters are enabled with a denoise coder.

6.2.1 Energy Consumption

A two-way radio communication should be preferred between the transmitter and receiver for transmitting the packets in such a way that the single-hop energy always remains identical to the energy that is transmitted, as shown in Equation 6.1:

$$\text{Sin gle-hop energy } (\text{SH}_e) = \text{Transmission energy } (\text{TE}_e), \tag{6.1}$$

where:

$$\text{Transmission energy } (\text{TE}_e) = \left(\Delta E_{\text{elec}}\left(S\right) + \Delta E_{\text{amp}}\right) * p * l^2, \tag{6.2}$$

where ΔE denotes the energy for transmitting and receiving a one-bit message; ΔE_{amp} represents the amplified energy; p is the dimension of the packet; and l represents the distance between the two nodes.

Conferring to the standard (IEEE 802.15.6), it is indicated that a maximum of two-hop communication can be allowed in wireless body sensor networks. As a result, a multihop network is deliberated that allows a two-hop transmission between the transmitter and receiver, as indicated by the following equation:

$$\text{Multihop energy } (\text{ME}_e) = p * h \left| \text{TE}_e + \Delta E_{\text{data}} + \Delta E_{\text{rec}} * h - 1/h \right|, \tag{6.3}$$

where h denotes the number of hops per second; ΔE_{data} represents the energy related to data aggregation; and ΔE_{rec} represents the energy related to data reception.

If all the paths have analogous hops, then nodes are assigned with priorities. The reason behind this is the low energy dissipation. Therefore, the multihop relationship can be expressed as:

$$Z(k) = \text{EC}(k), \quad \text{where } k = 1, 2,..i. \tag{6.4}$$

The total energy for transmitting the data from the sender end is written as:

$$\text{Total energy} = \sum \text{TE}_e, \Delta E_{\text{rec}}, \Delta E_{\text{sal}}, \Delta E_{\text{CA}} \tag{6.5}$$

where ΔE_{sal} represents the energy during transmission of salutation, and ΔE_{CA} represents the energy related to channel access procedure.

It is observed that, during the transmission of data packets, more energy is wasted as compared to during the recognition of signals. Also, there is path loss if the data do not reach the receiver (due to less signal strength) at a precise interval. The path loss depends on two factors: (i) frequency of transmission, and (ii) distance between transmission and reception.

6.2.2 Path Loss

The path loss can be defined by the following equation:

$$\text{Path loss } (\Delta f, l)\ (db) = \text{Path loss } (0) + 10k \log_{10} l/l_0 + w, \tag{6.6}$$

where l embodies the distance between the transmitter and receiver; l_0 is the reference distance; Path loss(0) signifies the reference distance; and w exemplifies the scattering parameter.

The reference distance (Path loss(0)) can be calculated from Equation 6.7:

$$\text{Path loss}\,(0) = 10\log_{10}\left[\frac{4\pi\Delta fl}{c}\right] \tag{6.7}$$

The signal thus transmitted gets affected if the human body possesses different parameters. Therefore, the cos t (Equation 6.8) can be calculated by considering the energy and distance between each node:

$$\text{Cos t}\,(k) = {}^{\text{length}(k)}\!\big/_{\text{energy}(k)}, \text{ where } k = 1, 2, ..i. \tag{6.8}$$

The consumption of energy and rump routing can be expressed with minimization of cost, such that:

$$\text{Min Cost} = \sqrt{\sum_{k=1}^{n}\left(\text{energy}_{ki} - \text{averge energy}\right)^2 \Big/ Z}, \tag{6.9}$$

where Z represents the number of available sensors.

6.2.3 Throughput

Throughput is demarcated as the aggregate quantity of effectively transmitted and received packets, where it is an obligatory constraint that the throughput of the intact network be exploited. Therefore, the optimization process for maximizing throughput is expressed as:

$$\text{Max throughput} = \sum_{i=1}^{n} \text{TR}_i\; i \in N, \tag{6.10}$$

where TR_i represents the number of successfully transmitted packets, and it entirely depends on network lifetime (N).

The necessity for transmission range is to discover the minimum number of sensors, which in turn examines each physical point with the help of a functioning measuring device, where a set of sensors are distributed over a large geographical area. For decreasing power consumption, a minimum set of working nodes are carefully chosen, which in turn makes the network lifetime longer:

$$\text{Min TR} = \sqrt{\left(\left({}^{S_{nD}}\Big/_{{}^{S_{n0}}/_{\Gamma_k}}\right)\right)}, \tag{6.11}$$

where S_{nd} and S_{n0} represent the degrees of preferred node and current node, respectively; and Γ_k denotes the coverage area that balances the extent that is concealed by the cluster.

6.2.4 Constraints

The network throughput as indicated in Equation 6.10 is subject to subsequent constraints:

$$\mathrm{TR}_i > \mathrm{TR}_k, \tag{6.12}$$

where Equation 6.11 specifies the dropped information packets when the transmission occurs between i and k.

$$\mathrm{ME_e} \geq \mathrm{TE_e} \tag{6.13}$$

The constraint indicated in Equation 6.12 shows that a data transmission has not yet transpired, as the transmitted energy is much lesser than the multihop energy.

6.2.5 Communication Distance

The communication of data packets between two nodes should be maximized in such a way that the distance of the first link is deliberated from the training node to the applicant node. Then, after estimating the distance of the first link, the detachment of the second link is recognized from the applicant node to sink node. Therefore, the distance stricture always guarantees the delivery of the packet to the destination node, which is given by the following equation:

$$\text{Min CD}\,(t, n) = \frac{C_1(t) + C_2(t)}{\sum_{n \in N} (C_1(n) + C_2(n))}. \tag{6.14}$$

6.3 Stacked Denoise Autoencoder

One of the most important types of deep learning techniques is the denoise coder, where all the corrupted versions of signals that are received by the receiver are recovered automatically [9]. The process is possible only by mapping both the input and output signals. It is necessary that a pretraining process be provided to the network for minimizing errors that happen between the transmission and reception of signals. In a denoise coder, the hidden layer always provides the output that needs to be predicted from the sequence of signal transmission. In the first step of the training process, input data are converted to a standard form for transmission. The entire range of the signal to be transmitted is obtained; therefore, any error that is present during the signal transmission process will be automatically decoded when SDAE is implemented. This type of deep learning technique is always necessary for allowing the signals to be passed regularly without any noise. For making the process easier, two steps have been designed—encoding and decoding. In the transmitter part, an encoder is present for mapping the input with the hidden layer, and, at the receiver part, a decoder is attached for retrieving the original signal.

One of the primary advantages of SDAE is that the convergence of the considered parameters, such as cost, lifetime, and energy, is faster than in other deep learning techniques. In addition, the ability to search for local optimum solutions is very much higher in SDAE. Figure 6.2 shows the implementation of a sensor with the SDAE process, where more sensors are placed at the transmitting end, which is referred as the input layer. The input layer is combined with an encoder, where any errors during sensor data processing are detected. Then, the input layer is combined with hidden layer.

The signals are sent through the hidden layer for correcting the errors. Then, at the receiver side, the hidden layers are combined with the output layer, where the signals are decoded to multiple sensors. In this process, multiple sensors can be integrated for the health-monitoring process by following the same procedure of encoders and decoders.

6.4 Comparative Analysis of SDAE

The health-monitoring system has been integrated with SDAE, and the results are simulated using MATLAB encoder toolbox. SDAE is one of the intelligent fault analysis techniques where the faults in the human body are analyzed accurately. SDAE is used in different

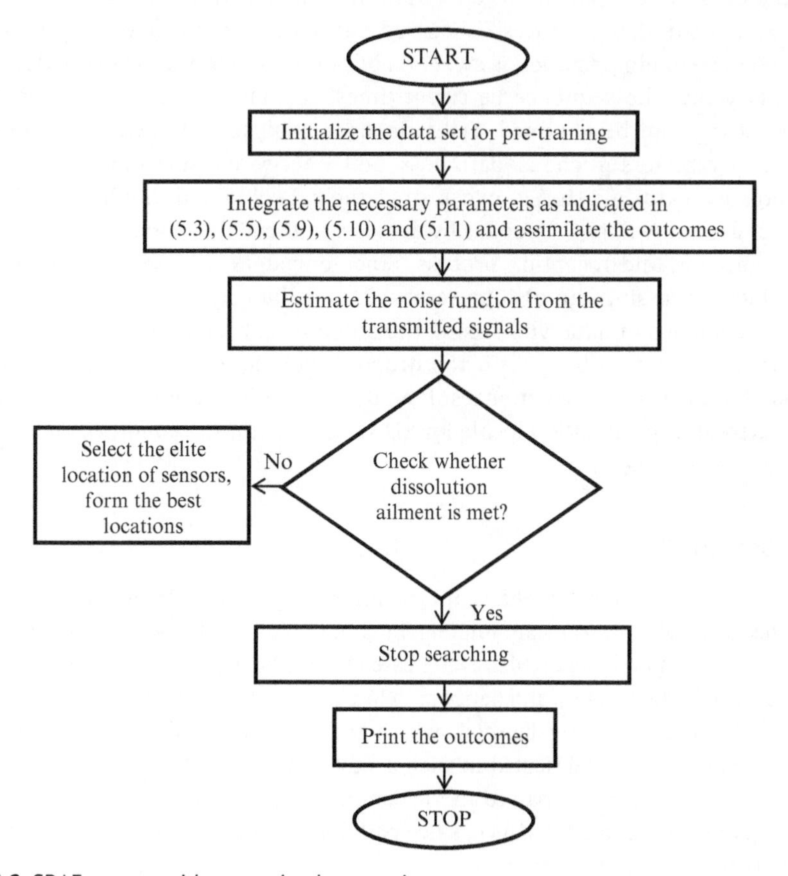

Figure 6.2 SDAE process with sensor implementation.

applications, but when it is applied for health monitoring, the vibrations are monitored with a particular frequency spectrum.

The simulation setup for health monitoring using SDAE has been executed, and similar works of other researchers that include artificial neural networks have been investigated, and the results are also compared. The results are simulated by considering parameters such as path loss, cost, energy, and lifetime. The results are also compared for these parameters using artificial neural networks. The reason for making comparisons with artificial neural network is that, when other methods are used, the error value rises in terms of both energy and accuracy. The experimental scenarios can be divided as follows:

- *Scenario 1*: Analysis of sequence behavior.
- *Scenario 2*: Investigation of path loss.
- *Scenario 3*: Determination of energy and throughput.
- *Scenario 4*: Calculation of network lifetime and cost.

6.4.1 Scenario 1

Since the data to be attained are for different behaviors such as static and dynamic positions, the simulation process is carried out with different frequency ranges. For each sequence, the information is switched between the array and sink nodes. Also, it is necessary that the sequence be run n times, so that an exact prediction for any human activity can be obtained. Daily data for analysis obtained from the human body for parameters such as path loss and energy consumption is substantial. Therefore, owing to data storage constraints, the data obtained should be limited. One way of reducing the data obtained is by decreasing the active nodes that are useful for the transmission and reception process. Since encoders are used, errors that occur during data processing can also be removed, and the original signals can be recovered. The sequence of data while using the autoencoder technique paves the way for monitoring the human body with the dropout technique, where the false data are rejected at both ends. For a better result analysis, SDAE is compared with neural networks where the specification locale for SDAE to monitor the condition of the human body is given in Table 6.1.

6.4.2 Scenario 2

To investigate the efficiency of the health-monitoring system, path loss needs to be determined, because, when sensors are implemented in the human body, there is a high probability of increase in path loss, which is indicated by decibel values. Similarly, the sequence also extends to higher ranges, and hence path loss increases in an extreme way. To decrease path loss, the deep learning method, which includes the denoise coder technique, is employed. The sensor is fabricated in such a way that the dipoles are integrated and the signals from the sensors are passed to the central node. The sensors and the dipoles are always separated by a small distance, since path loss increases rapidly if the distance between dipoles and sensors is very high.

Table 6.1 Specifications for health monitoring using SDAE.

S. no.	Parameters	Value
1	Area of simulation	0.8 m × 1.8 m × 0.3 m
2	Sensing node	16
3	Sensing nodes at front	8 (1–8)
4	Sensing nodes at side	4 (9–12)
5	Sensing nodes at rear	4 (13–16)
6	Thickness of body	30 cm
7	Position of sink node (m)	(0.4, 0.9, 0.3)
8	Node locations at front (m)	1 (0.4, 1.6, 0.3)
		2 (0.35, 0.95, 0.3)
		3 (0.3, 0.1, 0.3)
		4 (0.5, 0.4, 0.3)
		5 (0.5, 0.75, 0.3)
		6 (0.6, 1.05, 0.3)
		7 (0.55, 0.1, 0.3)
		8 (0.35, 1.6, 0.3)
9	Additional node locations (m)	9 (0.1, 0.9, 0.15)
		10 (0.7, 0.85, 0.15)
		11 (0.65, 1.05, 0.15)
		12 (0.7, 1.1, 0.15)
10	Rear side nodes (m)	13 (0.4, 1.25, 0)
		14 (0.55, 0.9, 0)
		15 (0.6, 1.1, 0)
		16 (0.6, 1.35, 0)
11	Reference distance	10 cm
12	LOS (η)	3
13	NLOS (η_1)	6
14	Frequency	2.4 GHz
15	Data packet size (k)	6000 bit

If artificial neural networks are used, then path loss is much higher, since more networks are present and the distance between the networks is also very much higher. Moreover, three layers are present in neural networks, which increase path loss to the maximum value. Therefore, an efficient method using SDAE is implemented with less distance separation. The other reason for increase in path loss is the presence of noise in the signals passing from the sensor to different nodes. But when SDAE is incorporated with sensors, the noise that is present in the signal is also filtered.

If the problems of both distance and noise are resolved, then perfect measurements can be obtained from the human body for extracting different channel scenarios, which even includes in-body channels. Figure 6.3 shows the comparison of path loss with different deep learning methods. The brown-dotted lines represent the SDAE method, where less path loss has been observed. The green, blue, and red lines represent the stacking autoencoder (SAE), long term short (LTS), and K-nearest neighbor methods [1], respectively. The existing method [9], which is indicated in rotund blue, shows that the neural network model is nearly equal to SDAE for the first half sequence. However, it has been observed that, if more sequences are present, the value of path loss is increased, which is indicated using a curved blue line.

6.4.3 Scenario 3

The data collected by sensors should be transmitted successfully, so as to fit within the exact values. The primary factor in data throughput is the energy parameter—that is, the transmitter side should have sufficient energy available for transmitting the signal to the receiver. For attaining sufficient energy, individual nodes at each sequence should be able to provide the necessary power. If adequate power is not provided, then the data obtained from the sensors cannot reach the receiver. The advantage of implementing SDAE for throughput analysis is that it has one additional layer, that is, the hidden layer. The function of this hidden layer is to block the signals that have very less energy; therefore, only the signals with higher energy are transmitted. The overall noise is also reduced in this case, as the number of transmitted signals is very less. The nonlinear characteristics of SDAE make the sensors track the signals that are obtained by the receiver.

The throughputs for artificial neural networks are also compared with SDAE where the neural networks cannot produce adequate throughput even with the presence of hidden layers. However, the deep learning technique with SDAE is capable of increasing the efficiency at all sensor locations. Figure 6.4 shows the comparison of throughput with other techniques where the violet color bar represents the SDAE deep learning technique.

It has been observed that, with a greater number of sequences, the denoise coder method performs well in terms of throughput. The blue, red, and green colors indicate the SAE, LTS, and K-nearest neighbor methods, respectively [1].

The violet color represents the neural network model where the throughput is much lesser in all sequences. Due to the separation of different layers from each other, the neural

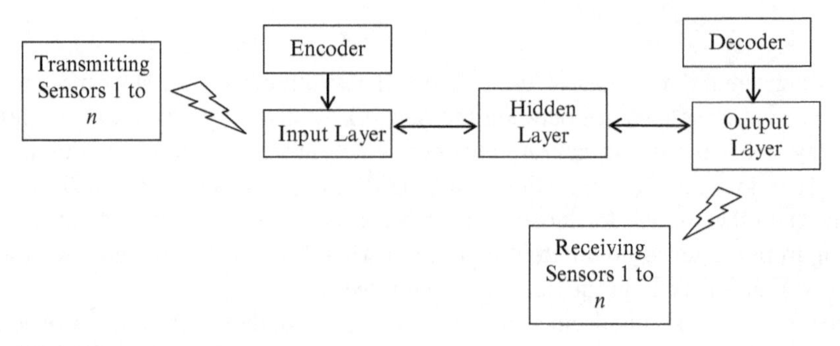

Figure 6.3 Path loss comparisons.

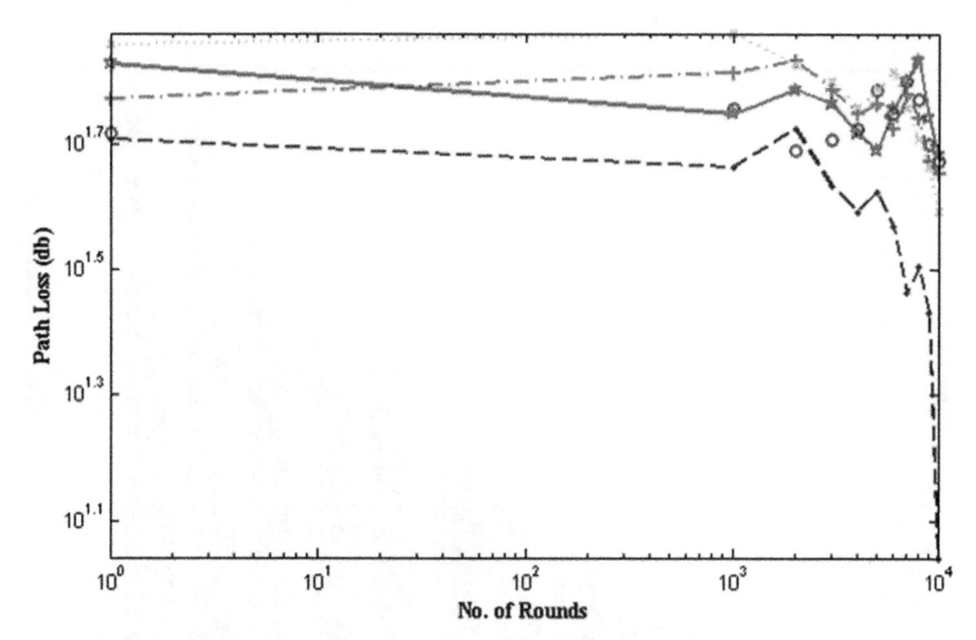

Figure 6.4 Throughput comparisons.

network model proves to be less efficient in terms of throughput. Figure 6.5 represents the energy consumption where the curved line in rose color indicates SDAE with a high consumption of energy. The SAE, LTS, and K-nearest neighbor methods [1] are represented using violet, rose, and green colors, respectively. It is observed that the neural network models, which are indicated in red color, have less energy consumption when compared to other techniques. Therefore, in terms of energy consumption too, the denoise coder proves to be more efficient.

6.4.4 Scenario 4

The pretraining cost of SDAE should always be lesser to achieve increased lifetime capability of the entire system using sensors. Therefore, it is important to design the loss function where it is calculated in total for SDAE and sensors. If the pretraining cost is kept within the limit, then all the users can shift to sensor-based technology at a cheaper cost. Here, the cost of each node should be reduced to a certain extent, so that the total cost can be automatically reduced. For achieving both cost and lifetime constraints, SDAE should be incorporated, because a cost-sensitive training is involved.

If accuracy is achieved, then the cost of the reliable networks can be decreased automatically with minimized reconstruction loss. In SDAE, the three layers can be incorporated as a single feature, whereas, in neural networks, the layers must be incorporated as a separate feature as more layers are present, which increases the cost of the entire network. Figure 6.6 represents the pretraining cost analysis, which is calculated using the number of nodes in the training network. For both 8 and 16 numbers of nodes, the denoise coding method

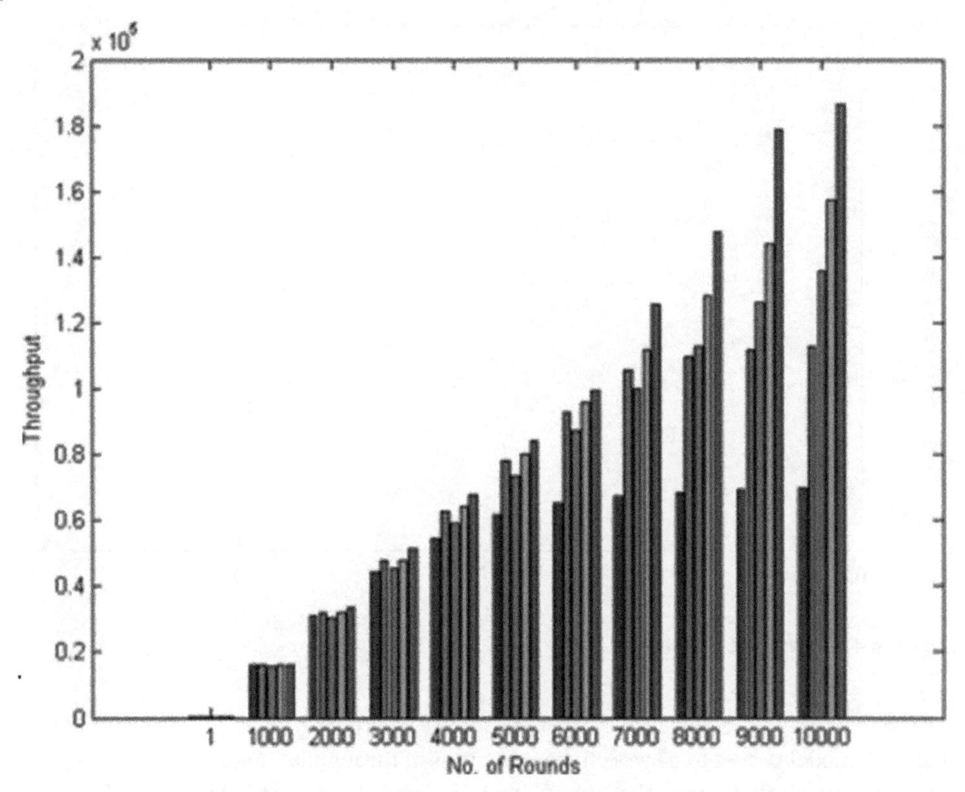

Figure 6.5 Consumption of energy.

produces less cost as compared to the other four methods. For 8 nodes, the total cost of SDAE is $1.2, whereas for 16 nodes, it is $1.5. But when neural networks are implemented, then, for 8 and 16 nodes, the cost is much higher.

The network lifetime is shown in Figure 6.7, where the violet color indicates the SDAE method. Here, within a lesser number of sequences, the lifetime of the entire network increases to a greater extent. But with increasing number of sequences, the lifetime of other methods, particularly for neural network, which is indicated in rose color, is the same as the lifetime of SDAE only after more number of sequences are distributed. Therefore, the SDAE deep learning method proves to be more efficient in terms of both cost and network lifetime.

6.5 Conclusion

For monitoring the health of individuals, body sensors with different sensor nodes were installed with the support of a special deep learning technique, that is, SDAE. Deep learning techniques have replaced the technologies that were based on standard algorithms. Therefore, the deep learning technique (SDAE) has established a new benchmark in

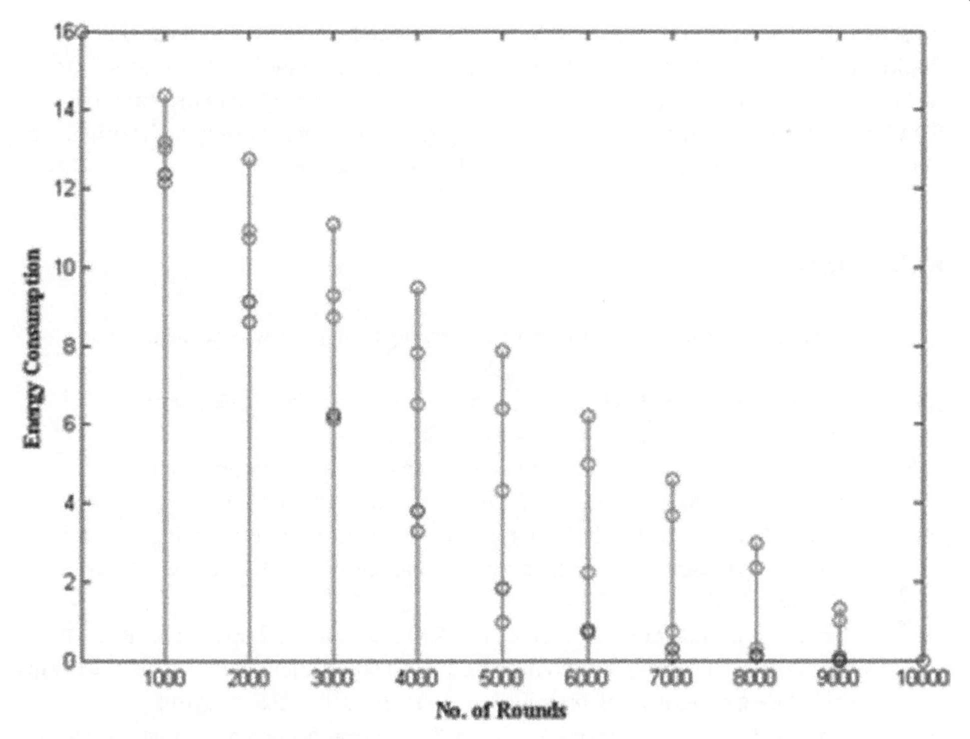

Figure 6.6 Evaluation of total cost ($).

sensor-based systems, particularly for the application of health monitoring, and it is able to bridge the gap that exists in neural networks. A comparison has also been included, where the SDAE technique proved much easier to implement than traditional neural networks, and it is possible to achieve very high performance with SDAE because, owing to the

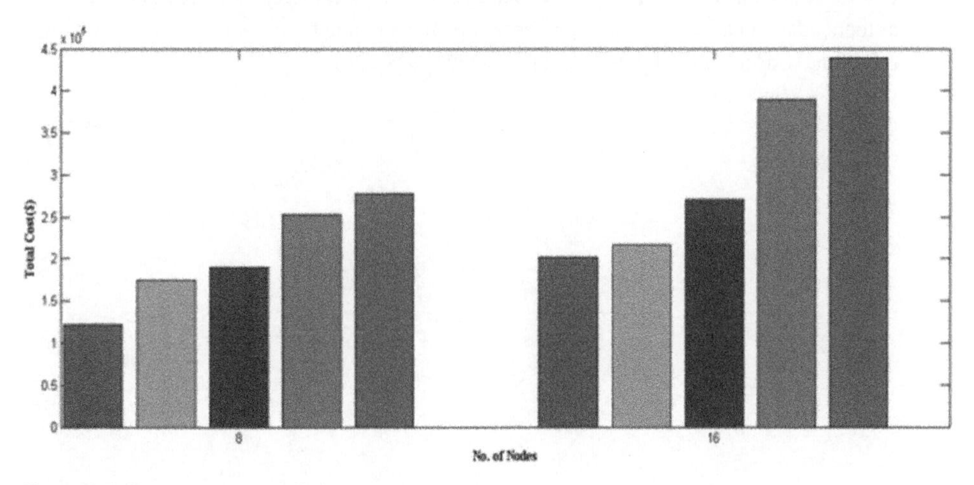

Figure 6.7 Overall network lifetime.

presence of encoder, the level of noise in the system can be tuned without causing any difficulties to the entire system. The efficiency of SDAE was also tested with four deep learning techniques, including SAE, LTS, K-nearest neighbor, and neural network methods, and SDAE proved very efficient in terms of sensor integration for monitoring the health of individuals when different sensor nodes are placed in the human body.

References

[1] S. Teotia and V. Sharma, "A review on deep learning models for wireless sensor networks," *Int. J. Sci. Res.*, vol. 8, no. 4, pp. 268–273, 2019.

[2] Z. Che, S. Purushotham, R. Khemani, and Y. Liu, "Distilling knowledge from deep networks with applications to healthcare domain," pp. 1–13, 2015.

[3] L. Xue, et al., "Understanding and improving deep neural network for activity recognition," *Int. Conf. Mob. Multimed. Commun.*, vol. 2018, June 2018.

[4] Z. Ali, G. Muhammad, and M.F. Alhamid, "An automatic health monitoring system for patients suffering from voice complications in smart cities," *IEEE Access*, vol. 5, no. c, pp. 3900–3908, 2017.

[5] L. Clifton, D.A. Clifton, M.A.F. Pimentel, P.J. Watkinson, and L. Tarassenko, "Predictive monitoring of mobile patients by combining clinical observations with data from wearable sensors," *IEEE J. Biomed. Heal. Informatics*, vol. 18, no. 3, pp. 722–730, 2014.

[6] M.S. Hossain, G. Muhammad, and A. Alamri, "Smart healthcare monitoring: A voice pathology detection paradigm for smart cities," *Multimed. Syst.*, vol. 25, no. 5, pp. 565–575, 2019.

[7] A.A. Obinikpo and B. Kantarci, "Big sensed data meets deep learning for smarter health care in smart cities," *J. Sens. Actuator Netw.*, vol. 6, no. 4, 2017.

[8] E. Kanjo, E.M.G. Younis, and C.S. Ang, "Deep learning analysis of mobile physiological, environmental and location sensor data for emotion detection," *Inf. Fusion*, vol. 49, pp. 46–56, 2019.

[9] P. Vincent, H. Larochelle, I. Lajoie, Y. Bengio, and P.A. Manzagol, "Stacked denoising autoencoders: Learning useful representations in a deep network with a local denoising criterion," *J. Mach. Learn. Res.*, vol. 11, pp. 3371–3408, 2010.

7

A Quantum-Behaved Particle-Swarm-Optimization-Based KNN Classifier for Improving WSN Lifetime

Ajmi Nader, Helali Abdelhamid, and Mghaieth Ridha

Micro-Optoelectronic and Nanostructures Laboratory, University of Monastir, Faculty of Sciences of Monastir, Tunisia

7.1 Introduction

Currently, the new technology called Internet of things (IoT) is seen to be successful in many fields. This technology is the result of the development and combination of different technologies where wireless sensor networks (WSNs) [1] are recognized as key enablers. The use of WSNs is increasing day by day and has gained a lot of research interest in various practical applications in domains such as healthcare [2], military defense [3], environment [4], monitoring [5], and industries [6]. A typical sensor network contains a set of tiny devices that are deployed in a two-dimensional area in a random manner to monitor a specific phenomenon such as humidity, temperature, motion, vibration, pressure, etc. The general architecture of WSN is shown in Figure 7.1. Some of the constraints of WSNs include limited communication range, lower data rates, and higher energy consumption. But the main challenge with WSNs is to extend the network lifetime.

There are many previous techniques proposed in different ways to resolve this issue. Among them, classification algorithms are one of the best solutions, where the k-nearest neighbor (KNN) algorithm [7] is the most widely used for classification in research.

In WSNs, another component that plays a very important role and has a high impact on network performance because of its advantages is called a *base station* (BS) or a *sink*. It has much more energy than all the other sensor nodes in the network, and is capable of calculating all the information regarding the distance, residual energy, and position of each node. This component acts as an intermediary between sensor nodes and the user to collect and forward data. Hence, a new challenge in WSNs involves searching for an efficient solution for selecting the optimal BS location. A lot of studies assume that optimal location of the BS can reduce long-distance communication and therefore minimize energy consumption.

On the other hand, there are several optimization algorithms that have been successfully applied in WSNs to improve both the network energy and lifetime. The most

Sensor Data Analysis and Management: The Role of Deep Learning, First Edition. Edited by A. Suresh, R. Udendhran, and M.S. Irfan Ahmed.
© 2021 John Wiley & Sons, Ltd. Published 2021 by John Wiley & Sons, Ltd.

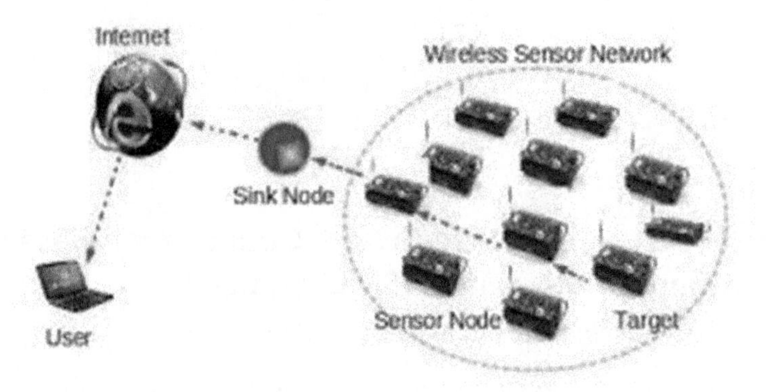

Figure 7.1 The general architecture of WSNs.

well-known among these are the PSO algorithm, genetic algorithm (GA), ant colony optimization (ACO), artificial bee colony (ABC), etc. On the other hand, some other studies combine two optimization algorithms for improving WSN lifetime, as seen in Reference [8].

In this chapter, we consider BS location as a solution to extend network lifetime in WSNs, where optimal location of the BS can reduce long-distance communication and therefore also minimize energy consumption. All this helps in extending network lifetime. In this context, a novel approach based on the hybridization of KNN and quantum-behaved particle swarm optimization (QPSO) is adopted in this work.

- A novel approach based on a KNN classifier algorithm and QPSO is used to identify an energy-efficient, optimal BS location.
- QPSO is applied to optimize KNN with a fitness function that considers two parameters—energy and distance.
- The communication distance between BS and nodes is minimized by choosing the optimal BS, so that less energy gets consumed, thereby prolonging network lifetime.
- We have evaluated the performance of the proposed algorithm using some of the performance parameters such as number of KNNs for BS, energy consumption, and network lifetime.
- The simulation is designed in scenarios depending on six different sizes of the network by using MATLAB.
- The experimental results clearly indicate better performance of QPSO-KNN over the well-known PSO-KNN algorithm in some important parameters.

The rest of this chapter is structured as follows. In Section 7.2, we address some of the recent algorithms in the existing literature related to our work. In Section 7.3, we present our network model and define the methods used in this chapter. Subsequently, Section 7.4 discusses the proposed algorithm in detail. In Section 7.5, we discuss the implementation details, and we conclude the chapter in Section 7.6.

7.2 Related Work

This section presents some of the previous studies that used swarm intelligence methods for improving WSN lifetime. Mohame et al. [9] adopted the PSO algorithm for selecting the efficient location of the BS within a special topology. They developed a fitness function to optimize some of the parameters such as energy of neighbors, number of neighbors, and distance between the neighbors and the middle of the network. The authors showed that the proposed algorithm has the potential to extend network lifetime in WSNs. Bogdanov et al. [10] proposed a method to build energy-efficient networks by determining the best location of the BS for maximizing the weight of data flows. In Reference [11], the authors addressed the objective of maximizing lifetime in WSNs. They used a strategy based on two methods, heuristic and greedy, in order to maximize the number of active nodes that support connectivity in the whole network. Meanwhile, some researchers proposed a corresponding strategy that combines two techniques to enhance WSN lifetime. Rana and Zaveri [8] proposed a method for optimization routing by using GA and PSO. The main idea of their work was to enhance the routes possible for gathering data. The fitness function aims to maximize the residual energy in the network. Another combination is proposed in Reference [12], where the authors suggest a method to enhance WSN lifetime by combining the KNN algorithm and whale optimization algorithm (WOA) [13]. The goal of this combination is to determine the best BS location that helps to save more energy in the network. The authors show that the suggested method improves energy consumption by 11%. In Reference [14], a PSO algorithm is adopted as a solution for the clustering technique problem in WSNs. The proposed algorithm aimed to divide the network into an efficient members of clusters and choose the best number of nodes as cluster heads.

7.3 Preliminaries

7.3.1 Network Model

Recently, many researchers proposed WSNs with specific characteristics according to situational requirements. In our case, we assume that WSNs consist of a large number of nodes as active nodes and inactive nodes, which are deployed randomly in a two-dimensional area. We deploy one BS or sink far from the sensor nodes. The BS is responsible for collecting and saving all data about sensor nodes in the network. Figure 7.2 shows our network model.

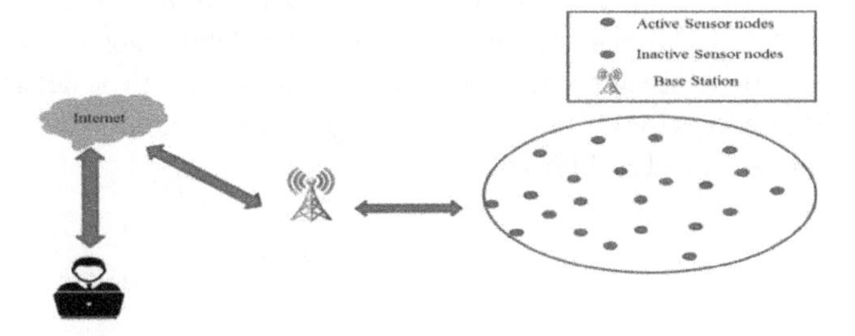

Figure 7.2 Network model.

We have considered some other assumptions about our network model, which are as follows:

- The BS is capable of communication with each sensor node in the network.
- All sensor nodes are homogeneous; they are capable of mobility and have the same initial energy $e_i > 0$.
- Each sensor node has a specific identification.

7.3.2 Quantum-Behaved Particle Swarm Optimization

A new edition of PSO is QPSO based on quantum theory. This technique is a kind of probabilistic algorithm, suggested by Sun et al. [15, 16], to avoid falling into the trap of local optima, which is often considered the most significant drawback of the conventional PSO technique [17, 18]. QPSO is considered as a global convergent method due an exponential distribution of positions [19, 20]. At variance with the canonical PSO, in QPSO algorithm, either (1) the particles of the population exchange their exposed position with their colleagues, or (2) the particle with the best position appeals to other particles to go for the optimal solution [19].

An important property of PSO [18] is that the accuracy of the algorithm is reached if the ith particle converges across its local optimum $\boldsymbol{p}_k^i := (p_k^{i,1}, p_k^{i,2}, \ldots, p_k^{i,d})^T$, which is defined as:

$$\boldsymbol{p}_k^i = \frac{c_1 \boldsymbol{P}_k^i + c_1 \boldsymbol{P}_k^g}{c_1 + c_2} = \varphi \boldsymbol{P}_k^i + (1-\varphi)\boldsymbol{P}_k^g \quad \forall (i,k) \in \left\{1, n_p\right\} \quad \text{Where } \varphi = \frac{c_1 r_{1,k}^i}{c_1 r_{1,k}^i + c_2 r_{2,k}^i}. \quad (7.1)$$

With reference to Equation (7.1), $r_{1,k}^i$ and $r_{2,k}^i$ are random numbers uniform over the closed interval [0,1], and c_1 and c_2 are the cognitive and social scaling factors.

In the quantum variant of the PSO algorithm, the population is a quantum system, and each particle has a quantum attitude, with its status determined by a wave function that must satisfy the Schrödinger equation [16, 18]. The new position of each particle can be achieved by utilizing the concept of Monte Carlo, which is described by the stochastic equation given in the following text:

$$\boldsymbol{x}_{k+1}^i = \boldsymbol{p}_k^i \pm \frac{\boldsymbol{L}_k^i}{2} \ln(1/\boldsymbol{u}_k^i), \quad (7.2)$$

where $\boldsymbol{u}_k^i$ is a uniform random number with standard uniform distribution on $(0,1)$, and $\boldsymbol{L}_k^i = (L_k^{i,1}, L_k^{i,2}, \ldots, L_k^{i,d})^T$ is a standard deviation of the distribution that defines, at each iteration k, the particle's search space. This standard deviation can be determined by calibrating the distance from the current position to a general best point $\boldsymbol{M}_k^{\text{best}}$, so the standard deviation can be calculated as follows:

$$\boldsymbol{L}_k^i = 2\rho \left| \boldsymbol{M}_k^{\text{best}} - \boldsymbol{x}_k^i \right|, \quad (7.3)$$

where the global point $\boldsymbol{M}_k^{\text{best}}$, called as the *mean best*, is determined by the average of the local best position among whole particles, given as:

$$\boldsymbol{M}_k^{\text{best}} = (M_k^{\text{best},1}, M_k^{\text{best},2}, \ldots, M_k^{\text{best},d}) = \left(\frac{1}{n_p} \sum_{i=1}^{n_p} P_k^{i,1}, \frac{1}{n_p} \sum_{i=1}^{n_p} P_k^{i,1}, \ldots, \frac{1}{n_p} \sum_{i=1}^{n_p} P_k^{i,d} \right) \quad (7.4)$$

Substituting Equation (7.3) into Equation (7.2), the equation of particle position can be written as:

$$\boldsymbol{x}_{k+1}^i = \boldsymbol{p}_k^i \pm \rho \left| \boldsymbol{M}_k^{\text{best}} - \boldsymbol{x}_k^i \right| \ln(1/\boldsymbol{u}_k^i) \quad (7.5)$$

where ρ is the unique parameter in quantum PSO algorithm, named *contraction–expansion coefficient*, and is used to adjust the convergence speed of the algorithm. Generally, the coefficient ρ decreases linearly during the search process from 1 to 0.5, but several works have shown that, when set to $\rho = 0.75$, the proposed algorithm can get a good performance with an acceptable convergence speed [16, 19].

Finally, the implementation steps of the QPSO algorithm for a minimization problem are outlined as the following pseudocodes [19, 21].

Algorithm 1. QPSO

1. Initialize the population size n_p and the particle positions $\boldsymbol{x}_0^i$, and determine the contraction–expansion coefficient ρ.
2. Calculate the mean best position vector $\boldsymbol{M}_k^{\text{best}}$ according to Equation (7.4).
3. Evaluate the corresponding fitness values $f_k^i = f(\boldsymbol{x}_k^i)$:

   (i) If $f_k^i \le \mathbf{pbest}_k^i$, then $\mathbf{pbest}_k^i = f_k^i$ and $\boldsymbol{p}_k^i = \boldsymbol{x}_k^i$;
   (ii) if $f_k^i \le \mathbf{gbest}_k$, then $\mathbf{gbest}_k = f_k^i$ and $\boldsymbol{p}_k^g = \boldsymbol{x}_k^i$,

where $\mathbf{pbest}_k^i$ and $\mathbf{gbest}_k$ represent the best previous fitness of the *i*th particle and the entire swarm, respectively.

4. Update the position of the particles using Equation (7.5):
   If rand(0, 1) > 0.5, then $\boldsymbol{x}_{k+1}^i = \boldsymbol{x}_k^i + \rho \left| \boldsymbol{M}_k^{\text{best}} - \boldsymbol{x}_k^i \right| \ln(1/u)$,
   else $\boldsymbol{x}_{k+1}^i = \boldsymbol{p}_k^i - \rho \left| \boldsymbol{M}_k^{\text{best}} - \boldsymbol{x}_k^i \right| \ln(1/u)$.
5. If the termination criterion is satisfied, the algorithm terminates with the solution $\boldsymbol{x}^* = \arg\min_{\boldsymbol{x}_k^i} \left\{ f(\boldsymbol{x}_k^i), \forall i, k \right\}$.

7.3.3 K-Nearest Neighbor

In the concept of feature selection and the classification problem, KNN [7] is most widely used for classification, as it is easy to implement, and training can done faster. KNN has been successfully used in different fields such as artificial intelligence and IoT security.

KNN is an intuitive supervised classification approach, often used in the context of machine learning. The principle of this model is mentioned in Reference [22], which is based on the choice of k data that have a short distance to the reference studied for predicting its value. The aim of this method is to search the KNN relative to the reference point. In this approach, some distance-calculating functions can be used, such as Euclidean distance, Manhattan distance, Minkowski distance, Jaccard distance, and Hamming distance. We used the brute force algorithm designed to search the KNN, as follows.

Algorithm 2. Brute force KNN
Input: Q, a set of query points, and R, a set of reference points.
Output: A list of K reference points for each query point.

1. For each query point $q \in Q$.
2. Compute distances between q and all $r \in R$.
3. Sort the computed distances.
4. Select K-nearest reference points corresponding to K-smallest distances.

7.4 Proposed Algorithm

In this study, we propose a QPSO-KNN-based solution to find the optimal location of the BS. The communication between the BS and nodes with short distances can save more energy in the network. Therefore, a node becomes able to live more, which makes the network more efficient in communication. So, the choice of location of the BS has a huge impact on the performance of WSNs. In this context, we adapted QPSO to optimize KNN to determine the optimal BS location. The aim of this approach is to improve the KNN classifier by selecting the optimal KNNs for the BS that is selected by QPSO using the fitness function. At the beginning of the algorithm, an initial population is created randomly that is composed of active and inactive nodes. The next step is to validate the KNN algorithm with the maximum iterations. The following step is to call the QPSO procedure for each iteration. In QPSO procedure, after this step, the fitness function can compute the population of N particles. Each particle has a value in the interval [0,1]. Then, the fitness function is used to judge the quality of each particle in the population. The cycle of QPSO procedure is explained in Section 7.3.2. The cycle continues until a termination criterion is satisfied. For this approach, the algorithm terminates with a list of the best values in the interval [0,1]. Finally, the KNNs keep all weights for classification. To clarify, after maximum iterations, a list of best solutions is presented, which is used in the KNN. Algorithm 3 gives a view of the proposed QPSO-KNN.

Algorithm 3. QPSO-KNN

1. Initialize network population.
2. While t is equal or less than maximum iteration:Apply update positions.Apply the evaluation of fitness function $f(x)$.Apply the selection of sink node location with the most convenient KNNs.
3. Choose the best solution.
4. Otherwise, go to step 2.

After selecting the optimal KNNs using the KNN classifier, another target of this study is to build the fitness function during implementation of the QPSO algorithm that considers the number of parameters for selecting the BS, such as neighbors' residual energy and distance between the neighbors and the middle of network. Our fitness function is given in Equations (7.6) and (7.7):

$$\text{Fitness function} = (\alpha_i * F(p_i), \qquad \forall\ F(p_i) \in \{D, E\}, \tag{7.6}$$

$$\begin{aligned} F(p_i) &= \alpha_1 D_p + \alpha_2 \sum_{i=1}^{N_p} E_{pi}, \\ F(p_i) &= \alpha_1 D_p + \alpha_2 \sum_{i=1}^{N_p} E_{pi}, \end{aligned} \tag{7.7}$$

where N_p is the number of neighbors, E_{pi} represents the residual energy of neighbors, and D_p is the Euclidean distance between the neighbor p and the center of network.

7.5 Performance Evaluation

In this chapter, we propose a novel approach to the optimal localization of the BS problem. The proposed approach, QPSO-KNN, combines two techniques. We used three parameters to evaluate the performances of QPSO-KNN, which are number of KNNs for BS and both the energy consumption and network lifetime. Finally, a comparison is made between the proposed approach and the well-known PSO-KNN. Our simulation model is discussed in the following text.

7.5.1 Simulation Parameters

To evaluate the proposed approach, we run six different sizes of WSN using Atarraya simulation [23], which is based on Java tools. The implementation of the proposed QPSO-KNN was done using MATLAB simulation, and the operating system was Ubuntu Server 16.04 Trusty. We used the energy model called *area-based collaborative sleeping protocol* (ACOS), which is defined in Reference [24]. The various simulation parameters that we used in our experimental scenarios are summarized in Table 7.1 and Figure 7.3.

7.5.2 Performance Parameters

In order to evaluate the proposed QPSO-KNN algorithm, we have used performance parameters such as number of KNNs for BS and both the network energy and lifetime.

1) *Number of KNNs for BS*: This parameter defines the total number of KNNs for the BS.
2) *Energy consumption*: This parameter is the total consumed energy of the transmitter data and receiver data along the network.
3) *Network lifetime*: This parameter defines the period from the beginning of communication in the network to its end (last node dies).

Table 7.1 Simulation parameters.

Parameter	Value
Deployment area	400 m × 400 m
No. of nodes	100, 200, 300, ... 600
Sensor node model	Simple
Node communication	Range 100 m
Node sensing	Range 20 m
Node location distribution	Uniform
Node energy distribution	Uniform
Maximum energy	1000 milliampere hour (mAh)
a_1	0.5
a_2	0.5

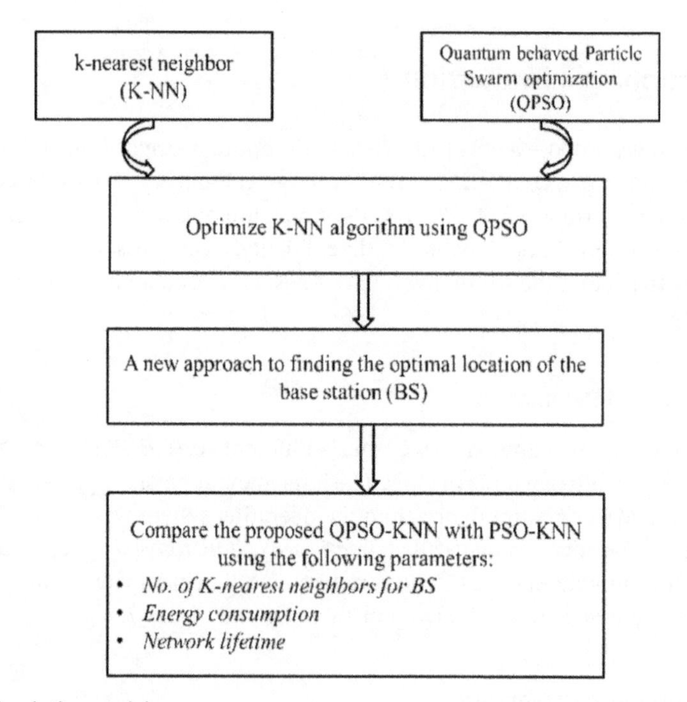

Figure 7.3 Simulation model.

7.5.3 Simulation Results

We run simulations by varying number of nodes (network size), and we compare our proposed QPSO-KNN algorithm with the well-known PSO-KNN algorithm in terms of number of KNNs for BS, energy consumption in the network, and lifetime.

Tables 7.2–7.4 show the results obtained in terms of the number of KNNs for BS, energy consumption, and network lifetime, which are presented in Figures 7.4–7.6, respectively.

1) *Number of KNNs*: It means the number of KNNs for the BS in both algorithms as shown in Figure 7.4.
2) *Energy consumption*: As seen in Figure 7.5, the proposed QPSO-KNN features an improvement in energy consumption, about 14% better than PSO-KNN. Table 7.3 displays energy consumption comparisons for both algorithms.
3) *Network lifetime*: Since positioning the BS at an optimal location of the network with the most convenient KNN may reduce the communication distance and save more energy consumption, the network lifetime increases. When a network size of 100 nodes is tested, the proposed QPSO-KNN shows a 23% improvement in network lifetime as compared to PSO-KNN, 16% in a network size of 200 nodes, 3.8% in a network size of 300 nodes, 8.2% in a network size of 400 nodes, 6.94% in a network size of 500 nodes, and 14% in a network size of 600 nodes. Figure 7.6 displays the network lifetime comparison between the proposed QPSO-KNN and the well-known PSO-KNN algorithm.

Table 7.2 Comparison of number of k-nearest neighbors for BS.

No. of nodes	PSO-KNN	QPSO-KNN
100	5	6
200	7	9
300	11	10
400	9	12
500	13	11
600	14	16

Table 7.3 Energy consumption comparison.

No. of nodes	PSO-KNN	QPSO-KNN
100	11 412	9436
200	13 604	11 083
300	13 862	12 576
400	16 233	14 403
500	14 564	13 789
600	18 285	16 034

Table 7.4 Network lifetime comparison.

No. of nodes	PSO-KNN	QPSO-KNN
100	1003	1234
200	1276	1481
300	1354	1406
400	1414	1531
500	1497	1601
600	1378	1572

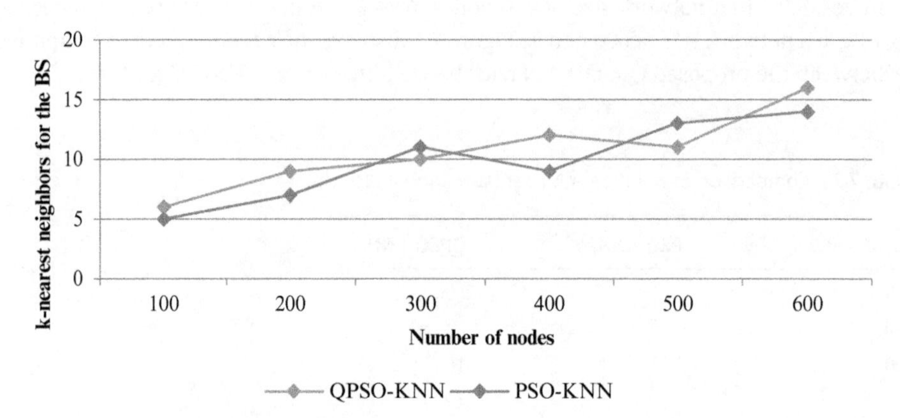

Figure 7.4 Number of k-nearest neighbors.

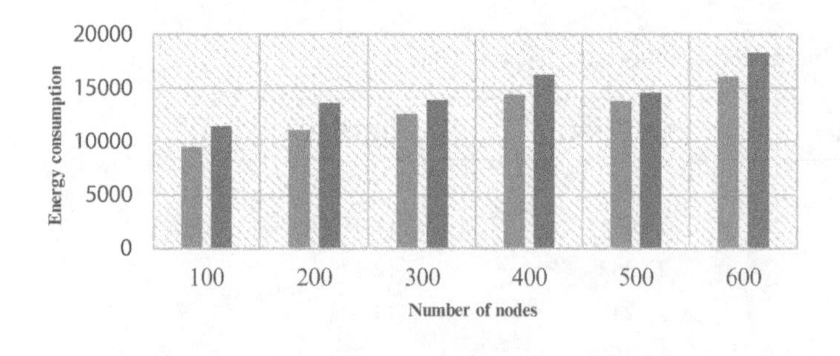

Figure 7.5 Energy consumption.

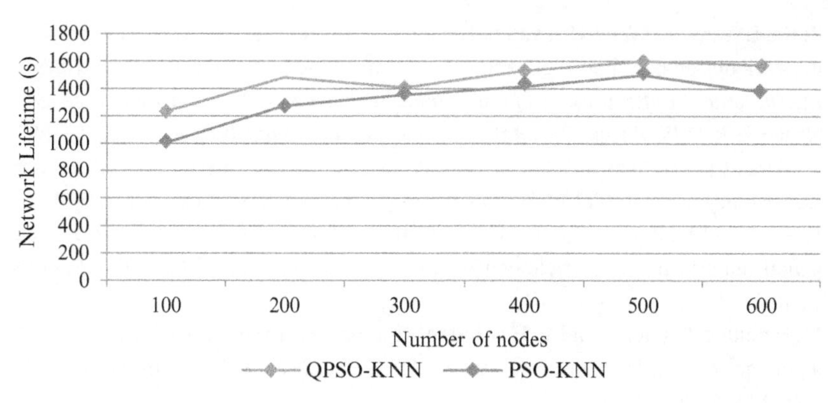

Figure 7.6 Network lifetime.

7.6 Conclusion

We proposed a hybridization of KNN and QPSO for improving WSN lifetime by updating the location of the BS in order to reduce long-distance communication between the BS and sensor nodes. QPSO is applied to optimize KNN. Furthermore, the fitness function is also designed considering two parameters—energy consumption and distances. The proposed QPSO-KNN is designed for better location of BS in the network with the most convenient nearest neighbors. The experimental results clearly indicate a better performance of QPSO-KNN over the well-known PSO-KNN algorithm in some important parameters such as energy consumption and network lifetime.

References

[1] J. Yick, B. Mukherjee, and D. Ghosal, "Wireless sensor network survey," *Comput. Netw.*, vol. 52, pp. 2292–2330, 2008.

[2] A. Moshaddique Al and J. Liu, "Security and privacy issues in wireless sensor networks for healthcare applications," *J. Med. Sys.*, vol. 36, pp. 93–101, 2012. doi:101007/s10916-0109449-4.

[3] M.P. Durisic, Z. Tafa, G. Dimic, and V. Milutinovic, "A survey of military applications of wireless sensor networks," In *2012 Mediterranean Conference on Embedded Computing (MECO 2012)*, Montenegro, 2012; pp. 196–199.

[4] M. Hammoudeh, F. Al-Fayez, H. Lloyd, R. Newman, B. Adebisi, A. Bounceur, and A. Abuarqoub, "A wireless sensor network border monitoring system: Deployment issues and routing protocols," *IEEE Sens. J.*, vol. 17, pp. 2572–2582, 2017. doi:10.1109/JSEN.2017.2674501.

[5] M.M. Hassan, B. Song, and E.-N. Huh, "A framework of sensor-cloud integration opportunities and challenges," In *Proceedings of 3rd International Conference Ubiquitous Information Management and Communication – ICUIMC'09*, Suwon Korea, 2009; p. 618.

[6] Z. Ke, L. Yang, X. Wang-Hui and S. Heejong, "The application of a wireless sensor network design based on ZigBee in petrochemical industry field," In *Proceedings of 1st International Conference on Computational Intelligence, Communication Systems and Networks ICINIS, Tetovo ,City in the Republic of Macedonia 2008*, 2008; pp. 284–287.

[7] E. Fix and J.L. Hodges, Jr "Discriminatory analysis-nonparametric discrimination: Consistency properties", California University Berkeley, Technical report, 1951.

[8] K. Rana and M. Zaveri, "Energy-efficient routing for wireless sensor network using genetic algorithm and particle swarm optimization techniques," *Int. J. Wireless Mob. Comp.*, vol. 6, pp. 392–406, 2013.

[9] M. Mohame, S. Fouad, and V. Hassanien, "Energy-aware sink node localization algorithm for wireless sensor networks," *Int. J. Distrib. Sens. Netw*, vol. 11, no. 7, 2015. doi:10.1155/2015/810356.

[10] A. Bogdanov, E. Maneva, and S. Riesenfeld, "Power-aware base station positioning for sensor networks," In *Conference of the IEEE Computer and Communications Societies*, Hong Kong China, 2004. doi:10.1109/INFCOM.2004.1354529.

[11] M. Cardei and J. Wu, "Energy-efficient coverage problems in wireless ad hoc sensor networks," *Comp. Commun.*, pp. 413–420, 2006.

[12] M. Mohammed, A. Ahmed, A. Taha, and H. Ehab "An optimized k-nearest neighbor algorithm for extending wireless sensor network lifetime," 2018. doi:10.1007/978-3-319-74690-6_50.

[13] S. Mirjalili and A. Lewis, "The whale optimization algorithm," *Adv. Eng. Softw.*, pp. 51–67, 2016. doi:10.1016/j.advengsoft.2016.01.008.

[14] R.K. Yadav, V. Kumar, and R.A. Kumar, "Discrete particle swarm optimization based clustering algorithm for wireless sensor networks," In *Advances in Intelligent Systems and Computing*, Springer International Publishing, Switzerland, V2, 2015.

[15] J. Sun, B. Feng, and W. Xu, "Particle swarm optimization with particles having quantum behavior", In *Proceedings of the IEEE Congress on Evolutionary Computation*, Portland, 2004; pp. 325–331. doi:10.1109/CEC.2004.1330875.

[16] J. Sun, B. Feng, and W. Xu," A global search strategy of quantum-behaved particle swarm optimization," In *Proceedings of the IEEE Conference on Cybernetics and Intelligent Systems*, Singapore, 2004; pp. 111–116.

[17] J. Sun, W. Xu, and B. Feng, "Quantum-behaved particle swarm optimization with a hybrid probability distribution", In *Proceedings of the Ninth Pacific Rim International Conference on Artificial Intelligence*, Guilin, China, 2006; pp. 737–746.

[18] J. Sun, C. Lai, W. Xu, Y. Ding, and Z. Chai, "A Modified quantum-behaved particle swarm optimization," In *Proceedings of the International Conference on Computational Science*, Kuala Lumpur, Malaysia, 2007; pp. 294–301. doi:10.1007/978-3-540-72-584-8-38.

[19] J. Sun, C.H. Lai and X.J. Wu, *Particle Swarm Optimization: Classical and Quantum Perspectives*, CRC Press, London, 2012.

[20] W. Fang, J. Sun, and W. Xu, "Improved quantum-behaved particle swarm optimization algorithm based on differential evolution operator and its application," *J. Sys. Simul.*, vol. 20, pp. 6740–6744, 2008.

[21] M. Xi, J. Sun, and W. Xu, "An improved quantum-behaved particle swarm optimization algorithm with weighted mean best position," *Appl. Math. Comp.*, vol. 205, pp. 751–759, 2008.

[22] O. Castillo, L. Xu, and S.L. Ao, *Trends in Intelligent Systems and Computer Engineering*, Springer, New York, 2008.
[23] M.A. Labrador and P.M. Wightman, *Topology Control in Wireless Sensor Networks: With a Companion Simulation Tool for Teaching and Research*, Springer Science& Business Media, New York, 2009.
[24] Y. Cai, M. Li, W. Shu, and M.-Y. Wu, "Acos: An area-based collaborative sleeping protocol for wireless sensor networks," *Ad Hoc Sens. Wirel. Netw.*, vol. 3, no. 1, pp. 77–97, 2007.

8

Feature Detection and Extraction Techniques for Sensor Data

Dr. L. Priya, Ms. A. Sathya, and Dr. S. Thanga Revathi

Department of IT, Rajalakshmi Engineering College, Chennai

8.1 Introduction

It has been a challenging task to ensure a reasonable quality of life in most countries. And now, achieving a reasonable quality of life has been greatly aided by the huge upgrades in medication and open social insurance. Subsequently, there is an immense interest in the advancement of dependable remote well-being observing [1], which could be anything but difficult to use for elderly people. Remote human well-being observing incorporates sensors, actuators, and propelled correspondence innovations and provides opportunities for the patient to remain at their home rather than in costly medical service facilities. This framework screens the physiological indications of the patients progressively, can make them stable under some well-being conditions, and passes this input to specialists. Why are these frameworks so agreeable and important to utilize? The primary explanation is that they are convenient and simple to use, with portable size and weight [2]. The primary model is a healthcare monitoring system (HMS) that has been designed using microcontrollers or embedded processors to track and send SMSs to a specialist's or caretaker's mobile phone, so that they can help in this scenario [3, 4].

The primary advantage of such setups is that the user is able to carry the system anywhere, anytime because of its portability. The secondary advantage is that their health conditions are being monitored continuously, that is, 24 × 7. People can use the system in hospitals, for remote care, and to track their vitals such as heart rate, blood pressure, and body temperature. These data can be processed by various sensors that are integrated into the systems [5, 6].

Effective high-speed microcontrollers or embedded processors are used in the design of health monitoring systems [7, 8]. The output of the system relies upon the number of wearable sensors or field programmable gate arrays (FPGA) kits as presented in Figure 8.1.

The heartbeat of the person is observed, processed, and communicated through wireless network for further follow-up. A centralized computer system is used for monitoring data transactions. The patient's data are observed and converted into an electrical signal and then displayed. Through this system, a person can wirelessly detect

Sensor Data Analysis and Management: The Role of Deep Learning, First Edition. Edited by A. Suresh, R. Udendhran, and M.S. Irfan Ahmed.
© 2021 John Wiley & Sons, Ltd. Published 2021 by John Wiley & Sons, Ltd.

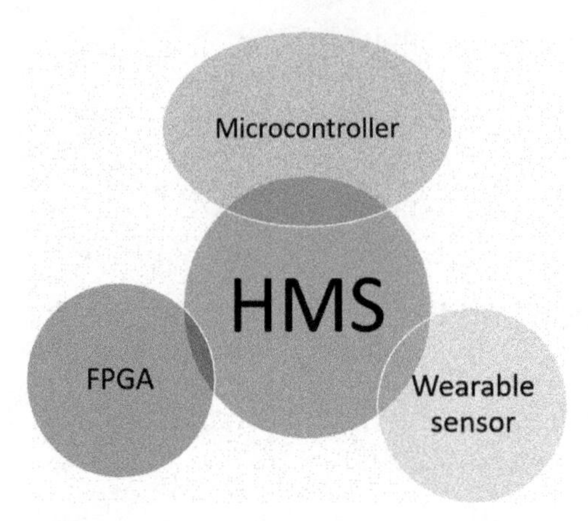

Figure 8.1 Health monitoring system.

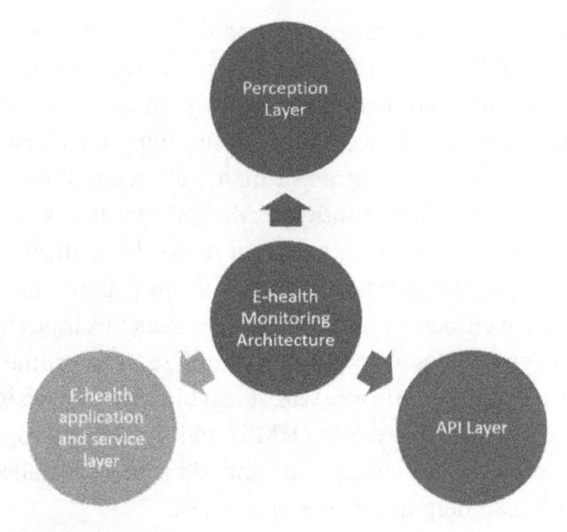

Figure 8.2 E-health monitoring architecture.

their own body temperature, heart rate, pulse, diabetic levels, etc. Standards such as Wi-Fi, Bluetooth, or any IEEE 802.11 wireless standard can be used for data communication [9, 10].

Sensors such as respiration sensor, electrodermal activity (EDA) sensor, and electromyography (EMG) sensor can be used as a wearable device for taking vital measures [11, 12]. A combined FPGA kit can be used for controlling the connected device. E-health monitoring architecture can be divided into three main layers as presented in Figure 8.2.

Multiple medical and environmental sensors are used to collect the data in real time through the perception layer. The patient's vital signs are measured through medical sensors, while environmental ones are measured through indicators. Environmental sensing could be done for identifying oxygen levels or room temperature for enabling a patient-friendly environment [13, 14].

The various programing applications are managed by the application programing interface (API) layer [15]. The data acquired through the perception layer are processed in the

API layer and stored in either private or public cloud or the patient's electronic health record (EHR). It stores new patient information and displays the existing medical information for registered users.

Data analysis and methods for improving the patient's condition are handled by the service layer. Data are analyzed by integrated algorithms and can be compared to other patient's experiences or previous health status of unique patient cases. In case of medical emergency, this layer is responsible for sending notifications to the physicians or staff who are connected in the network.

These kinds of systems are more compatible, can be configured easily because of their portability, and can save human lives.

8.1.1 Role of Sensors in Healthcare Systems

In general, a *sensor* is defined as a transducer that converts a measured quantity into processed data [16, 17]. Various types of signals include electrical, optical, and mechanical.

These devices are used to detect physical, chemical, and biological changes in the platform and provide a way for those signals to be measured and recorded. Vitals such as temperature, pressure, vibration, sound level, light intensity, weight, flow rate, etc., are observed. Sensor technologies that are utilized in current enterprise can be applied in medicine. In future, as particular sensors and sensor-dependent microelectromechanical structures are designed and tested, nonclinical industries will adapt them for commercial programs. In the global clinical scenario, there is a broadening intersection among information technology and biotechnology, and the position of sensors, sign transducers, actuators, and micromachines will become more prominent too. Some examples of latest-generation medical sensors indicate new roles that those gadgets will have in many areas of healthcare:

- A system for the continuous monitoring and recording of tissue glucose concentrations in diabetic patients (this has been approved in the United States). The subsequent generation of implanted glucose sensors will provide continuous monitoring values that can be read on far-away handheld devices or computers.
- A wearable device that looks like a watch worn on the wrist and produces small electric shocks that open up pores so that fluid can be extracted to monitor tissue glucose concentrations.
- Toto, a Japanese organization, has designed a toilet that analyzes urine for glucose concentrations, registers weight and different primary readings, and automatically sends an everyday record to the user's medical doctor with the aid of a modem.
- A closed-loop device (Scientific, San Diego, CA, USA) is now available for the blood evaluation of neonates in crucial care devices. Through an indwelling line, blood is circulated via a chemistry sensor, analyzed in a minute, and sent back, with no loss of blood to the neonate.
- Scientists have advanced a lightweight, completely automated machine for the detection of biological guns. This gadget uses fluorescent antibodies, diode lasers, fiber optics, and image detectors to discover airborne bacteria.
- Pathogens also can be detected with the aid of any biosensor that uses integrated optics, immunoassay strategies, and surface chemistry. Adjustments in a laser light transmitted with the aid of the sensor indicate the presence of a particular microorganism, and this record may be available within hours.

- An Australian crew has evolved a distinctly stable and touch-based biosensor that operates by means of switching the ion channels in a lipid membrane. While activated, biological receptors, which include antibodies and DNA, convert a chemical occasion into an electric-powered sign. One crew reports it could reduce the increase in sugar content material of dropping one sugar cube into Sydney harbor.
- The next generation of cardiac pacemakers will become "clever" by receiving readings from numerous places within the frame, consisting of oxygen saturation in the blood and cardiac wall pressure, permitting the pacemaker to align the coronary heart's pacing to the ones in real-time readings.
- Polymer wafers may be saturated with therapeutic markers, such as capsules or insulin. In the latter case, the wafer includes an enzyme that permits the wafer to change the pH and solubility of the insulin in reaction to modifications in blood glucose concentrations. Clever wafers being tested on animals use magnetism or ultrasound to adjust the dose of the drug or to replace it with another drug.
- Scientists and engineers at Johns Hopkins University have advanced a biochip photosensor that may be implanted as a synthetic retina for patients with macular degeneration and retinitis pigmentosa.
- Microorganisms causing ear, nostril, and throat infections may be instantly diagnosed by means of an electronic "nostril" that detects and differentiates the odors of the developing bacteria.
- At an international level, more than 20 000 human beings were implanted with auditory sensors that bypass the nonfunctioning elements of the hearing mechanism. The primary device that interfaces with the human brain is implantable and has a battery that is digitally programed and rechargeable through a transportable induction charging unit.

With advancements in technology, lots of smart or medical sensors have come into existence that can continuously analyze an individual patient's activity and automatically predict heart attacks before the patient feels sick. The following are the features of such smart or medical sensors:

- *Early warning for medical problems*: As per requirements, various types of sensors are available in the market that generate analytics to alert doctors. Before an adverse event occurs, these sensors can identify signs of heart failure, kidney failure, stroke, etc. For example, to detect cancer at an early stage, injectable biosensors are used.
- *Enable neural technologies*: Smart neural devices or sensors can prove to be an asset for patients to manage conditions such as rheumatoid arthritis or Parkinson's disease; they also integrate neural bypass technologies that are especially useful for paralyzed patients.
- *Automate smart medical devices*: Sensors are increasingly being used to automate smart medical devices. To fulfill real-time patient needs, various smart devices are created that combine data analytics and sensor data to adjust medication delivery.
- *Manage chronic conditions of patients*: Various tiny, high-resolution sensors allow patients and doctors to manage real-time chronic conditions such as heart disease, diabetes, multiple sclerosis, etc. To provide real-time recommendations for patients, various advanced analytical engines could monitor signals from multiple sensor types.

8.1.2 Types of Sensors in the Medical Field

In the medical field, patients nowadays actively take part in collecting and reviewing their reports. In this digitized world, various wireless communication standards have allowed sensors to develop from *traditional forms*, that is, those requiring active patient participation, to *passive forms*, that is, those that do not require patient participation.

Today, a large number of passive sensors are used that constantly monitor the vital signs of individual patients and store that data or share it wirelessly with healthcare professionals. By combining analytics and sensor data, reports are made that describe the early health condition of the patient. Depending on the requirement, various types of sensors are being deployed, such as follows:

- Wearables like virtual watches, Fitbit, and many others.
- Ingestible sensors embedded in capsules.
- Blood sampling sensors, which include glucose meters.
- Outside sensors, which include pulse oximeters, blood strain cuffs, etc.
- Epidermal sensors together with virtual tattoos and patches.
- Tissue-embedded sensors, including pacemakers and defibrillators.

In healthcare, different types of sensing devices are used depending upon their characteristics, usability, and efficiency. Their merits and demerits are as follows:

Merits:

- Easy to design.
- High reliability.
- Scalable bendy device.
- Minimum interconnecting cables.
- High overall performance.
- Small, rugged packing.

Demerits:

- In wired smart sensors, complexity is a lot better managed; as a result, their cost is also high.
- Predefined embedded characteristics are required during the design of smart sensors.
- They require actuators and sensors.
- Sensor calibration must be managed via an external processor.

The sensors used to diagnose, monitor, or treat diseases in medical domain are known as *medical sensors*. Medical devices are categorized into different classes based on their risk profiles, namely, Class-I (lowest potential risk), Class-IV (highest potential risk), etc. Let us understand the features and functions of different types of medical sensors. Figure 8.3 shows different sensors that can be placed in a human body.

The following are the functions of different types of medical sensors used for various applications:

- *Temperature probes*: They are used for body temperature measurement. This helps in providing better medication and treatment to patients. They are called *thermometers*.

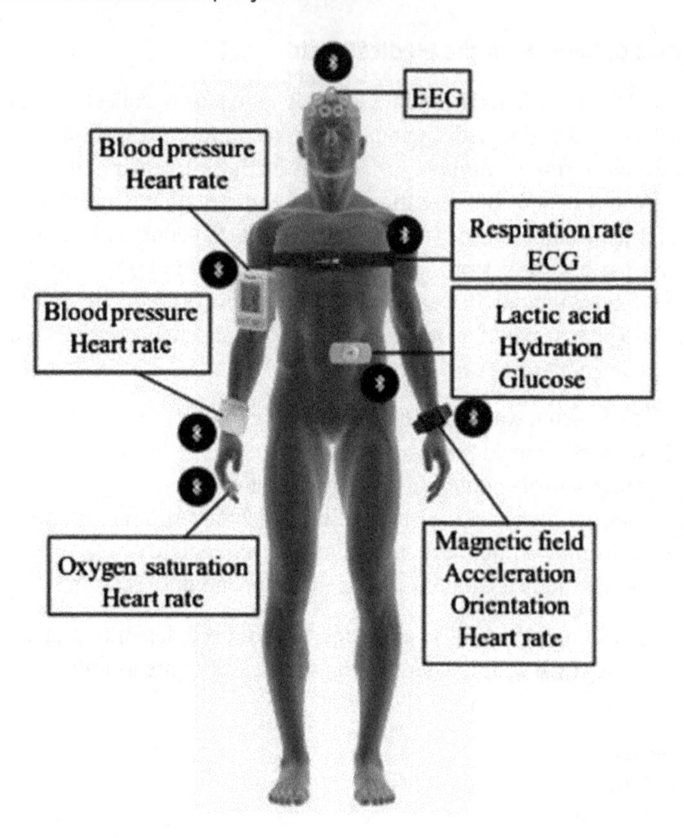

Figure 8.3 Sensors in human body.

- *Airflow sensors*: They are used in anesthesia delivery systems, laparoscopy, heart pumps, etc.
- *Force sensors*: They are used in kidney dialysis machines.
- *Implantable pacemaker*: It is a real-time embedded sensor system that delivers a synchronized rhythmic electric stimulus to the heart muscle in order to maintain effective cardiac rhythm.
- *Pressure sensors*: They are used in infusion pumps and sleep apnea machines. Most of the pressure sensors are integrated with embedded systems. They are used for medical diagnosis, blood pressure monitoring, infusion pumps, etc.
- *Magnetometer*: It specifies the direction of a user by examining the changes in the earth's magnetic field around the user.
- *Heart rate sensor*: It counts the number of heart contractions per minute.
- *Glucometer*: It measures approximate blood glucose concentration.
- *Oximeter*: It measures the fraction of oxygen-saturated hemoglobin relative to the total hemoglobin count in the blood.
- *Electrocardiogram sensor*: It measures the electrical activity of the heart. It is also known as *electrocardiography* (ECG) *sensor*.

- *Electromyogram sensor*: It records the electrical activity produced by skeletal muscles.
- *Respiration rate sensor*: It counts how many times the chest rises in a minute.
- *Electroencephalogram sensor*: It measures the electrical activity of the brain.

8.1.3 Research Questions

How can sensor data be processed?
How can features be extracted from sensors?
What are the parameters that can be extracted from sensors?

8.2 Related Works

Humans and computers have complementary strengths: human beings are precise at scanning large areas and spotting objects, while computer systems are precise at optimization, detailed delineation, and repetition. Whether manual, automated, or a combination of the two, feature extraction may be a totally involved method. Guide feature extraction harnesses the translation talents of the operator, but may be time consuming and, as a result, costly to carry out. With a growing body of digital data archived and a hastily changing society, the green revision of cartographic databases implies a few forms of automatic function extraction. The use of automated or semi-computerized strategies can also offer value savings by extensively lowering the education time of photograph interpreters.

Humans have the potential to organize easy features, consisting of factors and lines, into significant systems. Semi-computerized methods rely upon consumer-furnished cues to delineate road additives. Furthermore, the authors trust that the use of a method is most excellent due to the fact that people carry out identity almost perfectly with constrained effort. People are capable of understanding shapes in data and adapting to varying conditions, without being advised explicitly on what to expect. Writing computer code to simulate this capacity is a full-size project. Incorporating existing geographic information system (GIS) records into the feature extraction technique can lessen the need for human interference.

To carry out object popularity, it is important to first set up a or framework that describes the general characteristics of the feature of hobby. Computerized feature extraction requires the network defined in a way that can be applied via computer.

Totally model-based processing exploits the limitations and relationships that define gadgets, for example, the scale, shape, and cloth of a construction, or the width, fabric, and route of a road. The function version includes information referring to a range of characteristics consisting of intensity, shape, texture, and context. Models are frequently characterized as being both flexible and rigid. An inflexible version defines capabilities specifically, for example, outlining the allowable size, shape, and spectral reaction. A flexible version may consist of specs in phrases of usual constraints, together with smoothness, rectilinearity, curvature, compactness, symmetry, and homogeneity.

In the case of acceleration information, time-domain and frequency-domain capabilities are generally used, while time-based features along with a variety of coronary heartbeats are used on crucial signals. The dimension device is a metallic oxide gas sensor array containing five Figaro TGS collection SnO_2 sensors (TGS 800, 822, 824, 825 and 842) and modified to host a thermo-hygrometer and to accumulate the sensor voltage on a 12-bit A/D converter. Twenty-nine features are extracted from every sensor reaction curve consisting of seven suggested reaction values on one kind of time interval, six slopes on special time intervals, eight primary derivatives, and eight secondary derivatives. After this, these features are confirmed and evaluated qualitatively by using three particular indexes: repeatability, discrimination, and redundancy. However, the authors just proposed the three indexes to qualitatively examine the features, and did not give any classification outcomes for particular statistics regarding the use of special functions, which were considered higher than others in step with their proposed three evaluating indexes.

Further, the most important factor evaluation, principal component analysis (PCA), was used to analyze the features. The PCA result gave score plots and loading plots. The tendencies, agencies, and outliers of the samples may be located in rating plots, and the correlation and similarity between the functions can be found. An extended distance from the origin within the loading plot and a bigger sign-to-noise ratio (SNR) implied that the function contained greater facts. The synthetic neural networks (SNN) classification outcomes showed that: (i) it changed into feasible to prediction accuracy to use functions in step with the S:σ calculation; (ii) there are facts that could enhance recognition capacity in the loading plot; (iii) functions extracted from unique reaction curves may also gave similar consequences as the complex curve fitting coefficients.

A brand new approach is to detect wound pathogens by deciding on the wavelet rework coefficients preferentially with a scatter matrix and using the mean of the chosen coefficients as the function. The new function extraction method confirmed excessive performance in figuring out seven species of pathogens. Underneath the effect of a robust waft, even though each of the classification charges decreased, the anti-flow capability of the new capabilities turned obviously stronger.

8.2.1 How Can Sensor Data Be Processed?

Before extracting features from the dataset, it is essential to process the data obtained through sensors. The most common processing methods are related to data preprocessing, segmentation, and dimensionality reduction methods, as presented in Figure 8.4.

Dimensionality reduction operations include the extraction of those features that represent significant data characteristics and the posterior selection of specific discriminative features to reduce the dimensionality of the feature vector while maintaining most of the relevant information. The rationale behind these methods is related to the huge volume of raw data that may be generated in smart homes, due to the heterogeneity and ubiquity of the sensors used. The ability to extract summarized and useful information from raw sensor data becomes a key factor for the feasibility and performance of current smart home services.

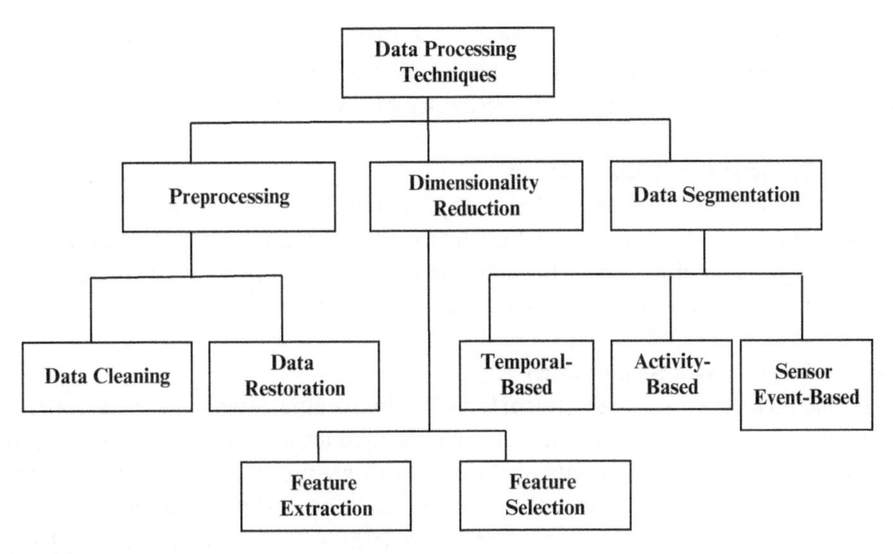

Figure 8.4 Various data processing techniques.

8.2.1.1 Feature Extraction

The main characteristics of original data are obtained from the quantitative measures and are represented as features. Extracting these kinds of representative features is helpful to improve the accuracy of results. These extracted features are helpful for further information-processing stages with less computational cost of activity inference. After extracting the features into a set of feature vectors, they are given as an input to the activity discrimination and learning algorithms. The most commonly used approaches of feature extraction operate in three domains: time domain, frequency domain, and discrete domain.

8.2.1.2 Feature Selection

The extracted data from raw sensors may contain redundant and irrelevant information, which can immeasurably affect system performance. Feature selection plays an effective role in selecting more discriminative features and reducing the dimensionality of feature vector. In this way, the feature selection process helps the system to find a more relevant subset of features with a high-dimensional feature vector. This, in turn, reduces the computational cost and expenses to benefit the application of learning models.

8.2.1.3 Data Segmentation

Information from sensors generally comes as a consistent progression of crude information. This is on the grounds that sensors will essentially give moment estimations of the checked marvel, either when mentioned or at occasional interims. One of the primary provokes identified with tangible information preprocessing is to accomplish a legitimate division of this crude and consistent information stream into smaller squares of data. This is the motivation behind division strategies. The best possible choice and parameterization of division systems have extraordinary potential effect on the achievement of highlight extraction and surmising calculations, straightforwardly bringing about the precision of movement observing and acknowledgment. Current writing features, for the most part, three classes of division draw—near fleeting-based division, movement-based division, and sensor occasion-based division, as clarified in the accompanying segments.

8.2.1.4 Temporal-based Segmentation

There exist two types of temporal-based segmentation approaches, namely, *time interval based* and *sliding window based*. The time-interval-based approach divides the sensor datasets into equal time durations. This is commonly used for breaking down the temporal stream data obtained by accelerometer and gyroscope sensors.

8.2.1.5 Activity-Based Segmentation

In this method, the sensory data stream is divided into multiple segments by identifying the start and end points of each activity. The main consequence of this method is the correct identification of boundary instants. Various methods are proposed to identify these limits, distinguishing between static activities (such as standing and sitting) and mobility (such as walking and running). In all approaches, a threshold is set for identifying the changing points of stand-alone activities and the analysis of changes in the frequency domain which is used for determining the beginning and ending points of movement-related activities.

8.2.1.6 Sensor-Event-based Segmentation

Sensor-occasion-based division approaches are utilized for perceiving exercises which comprise a grouping of developments, occasions, or activities that occur in a specific time request and that may be interleaved with other exercises' occasions, for instance, "family unit" or "dinner arrangement" exercises. Contrasted with fleeting-based division, right now, occasions that structure the action may not be disseminated consistently in time and may happen sporadically; in this manner, the size of the windows is not fixed.

8.2.2 How Can Features Be Extracted from Sensors?

In order to extract features, some degree of prior knowledge about the dataset is required. Datasets may contain multiple symbols, various motions, maxima and minima points, etc. For instance, motions like left turn, straight line, and right turn can be represented as a series of symbols, and domain experts may be used to retrieve or understand the given information. Quantifications like maxima and minima can be represented as a set of numeric values and may have compressed representation. Hence, extracting features from the dataset obtained through sensor outputs may require a set of processes, as shown in Figure 8.5.

The data collection unit collects the data from sensors, and it is compared with standard values of various sensors such as wearable sensors, accelerometers, pressure sensors, and few others. These sensors are centrally managed, and the data from these sensors are preprocessed such as unknown values, redundant values, and other well-known scenarios. The preprocessed data are further processed using standard signal processing techniques. The processed data features are extracted through feature extraction techniques such as data streaming and unsupervised learning. The extracted features may be time domain or frequency domain in nature. Time series features can be extracted for time series data, and feature extraction can be performed using various time series analysis and decomposition techniques.

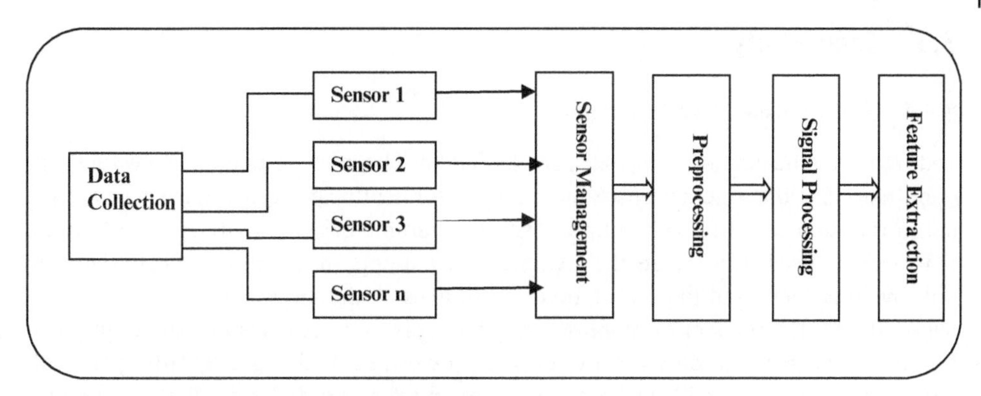

Figure 8.5 Feature extraction from sensor data.

Table 8.1 Sensor parameters and their descriptions.

Parameter	Description
Maximum response	Max (sensor value)
Integral	$I=\int_a^b f(x)\mathrm{d}x, I=\int_a^b f(x)$: sensor value. x: time from a to b; a: the beginning of the adsorption stage; b: the end of the desorption stage
Derivative	Difference between two samples during measurement
Polynomial function	$Y=A_0+A_1x+A_2x^2+A_3x^3$ On: $Y=$ (sensor value-baseline), $x=$ time from gas On to gas Off Off: Y = (final response-sensor value), x = time from gas Off to gas end
Fractional function	$Y=x/Ax+B$, where Y and x are defined as in the polynomial fit
Exponential function 1	$Y=A(1-exp(-x/T))$, where Y and x are defined as in the polynomial fit
Exponential function 2	$Y=A_0+A_1(1-\exp(-x/T_1))+A_2(1-\exp(-x/T_2)$. where Y and x are defined as in the polynomial fit
Arctangent function	$Y=A\times\arctan(x/B)$, where Y and x are defined as in the polynomial fit
Hyperbolic tangent function	$Y=A\times\arctan(x/B)$, where Y and x are defined as in the polynomial fit
Fourier descriptors	Coefficients of the fast Fourier transform (FFT) of sensor value
Wavelet descriptors	Coefficients of the discrete wavelet transform (DWT) of sensor value

8.2.3 Sensor Parameters

Brief descriptions of the parameters extracted from sensor data and their response curves are given in Table 8.1. While extracting features from sensor data, it is essential to understand and fit the following parameters into the responses for effective data analysis.

8.3 Case Study

8.3.1 Sensor-Based Health Monitor

Today, human healthcare using body sensor data has been implemented in a wide range of human–machine interactions. Smart systems for healthcare sectors have been getting more attention recently. For example, smart wearable-based activity recognition systems can be used to assist in the treatment of patients in a smart clinical system to improve treatment and to prolong their social lives. Although there are many ways of using distributed sensors to monitor the vital signs and activities of people, physical human action recognition via body sensors provides valuable data regarding an individual's functionality and lifestyle. In this chapter, we propose a sensor-based system for health monitoring through activity recognition using deep recurrent neural network (RNN), which is a promising deep learning algorithm that is completely based on sequential information. Data fusion has been performed using multiple body sensors such as ECG, accelerometer, magnetometer, etc. The extracted features are then enhanced via PCA, and the identified features are then used to train the neural network, which is later used for activity recognition. The system has been compared against conventional approaches on three publicly available standard datasets. The experimental results show that the proposed approach outperforms the available state-of-the-art methods.

We propose here a robust, deep RNN-based activity recognition system based on efficient features. After extracting the data from the different multimodal sensors, the data are further fused with statistical features of different orders such as mean, variance, standard deviation, skewness, and kurtosis. To make the features more robust, they are projected into a nonlinear space with dimension reduction process as presented in Figure 8.6. Kernel-based PCA is used to analyze data in the nonlinear space, which is better than typical linear feature space-based PCA. The robust features are then trained and tested via a memory-based RNN for activity recognition. To check and compare the proposed approach with the traditional ones, we chose three publicly available datasets and adopted different combinations of training and testing methods. The proposed approach shows superior performance on all the datasets. Pseudocode for the proposed approach is given in the following text:

```
SET training data
Preprocess the data
Build model
Get sensor data
        Perform preprocess and denoise data
        Build required dataset
Fix relevant features
Extract required features
```

Step 1: Load all training data in memory from a comma-separated values (CSV) file: every feature vector in the same line with the class at the end.
Step 2: Randomly divide the training data into *N* subsets (from 40 to 95 in this chapter).
Step 3: For every subset, a decision tree is built:
  Step 3.1: At the starting point, all feature vectors belong to a unique node.
  Step 3.2: At each node, all features and different thresholds are considered to split the node (using simple rules: one feature and one threshold).
  Step 3.3: The best split is selected according to an objective function (entropy gain).
  Step 3.3: If the best split improves node entropy over a threshold, the node is split; otherwise, the node is not split (stop criterion).
  Step 3.4: Steps 3.2, 3.3, and 3.4 must be repeated till a maximum number of iterations (100 in this chapter) is reached or when the nodes cannot be split further.

8.3.2 Case Study 2—Cloud-Based Mobile Healthcare

The biomedical signs contain numerous kinds of ancient rarities, for example, eye squinting, muscle, and other inside or outer meddling commotions, and these antiques ought to be cleaned. These ancient rarities can be evacuated by utilizing a few systems. The equipment channels existing in biomedical hardware can sift through a large portion of these ancient rarities and commotions. The biomedical signs are obtained and prepared to recognize distinctive client expectation designs. The biomedical sign investigation and handling are acknowledged in three principal steps: preprocessing/denoising, highlighting extraction/measurement decrease, and discovery/characterization. The fundamental objective of preprocessing is to rearrange succeeding methodology without losing related data and to improve sign quality by expanding the SNR. A lower SNR implies that the biomedical sign examples are stifled in the remainder of the sign, and the related examples cannot be recognized. Be that as it may, a higher SNR makes the grouping task more straightforward. Specialists utilize various strategies to dispose of or, if nothing else, decrease the undesirable sign segments by changing the signs. These techniques may improve the SNR. The proposed architecture is shown in Figure 8.7. The snippet for feature extraction is given in the following text:

```
FDR_LEVEL = 0.05X = extract_features(d_ts,
                    column_val='val',
column_sort='time',X_selected = select_features(X,Y)
print(X.shape)
print(X_selected.shape)

default_fc_parameters=ComprehensiveFCParameters(),
                     impute_function=impute)
x = x.loc[:, x.apply(pd.Series.nunique) !=1]df_pvalues_
mann=calculate_relevance_table(x, y, fdr_level=PDR_LEVEL,
test_for_binary_target_real_feature='mann')
```

```
print("Total\t", len(df_pvalues_mann))
Print("Relevant|t", (df_pvalues_mann[relevant"]==
True).sum())
print(Irrelevant\t", (df_pvalues_mann["relevant"]==
Fales).sum(),
   "(# constant", (df_pvalues_mannn["type"]=="const").sum(),")"
```

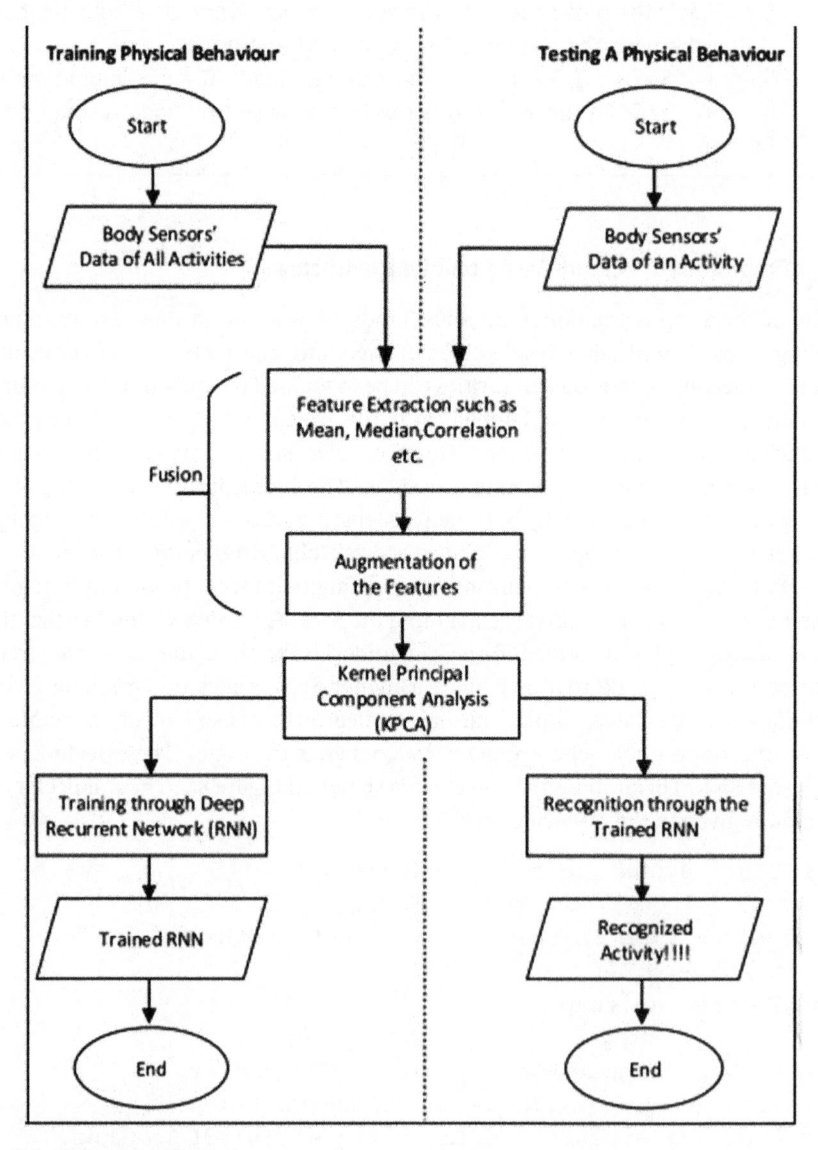

Figure 8.6 System architecture.

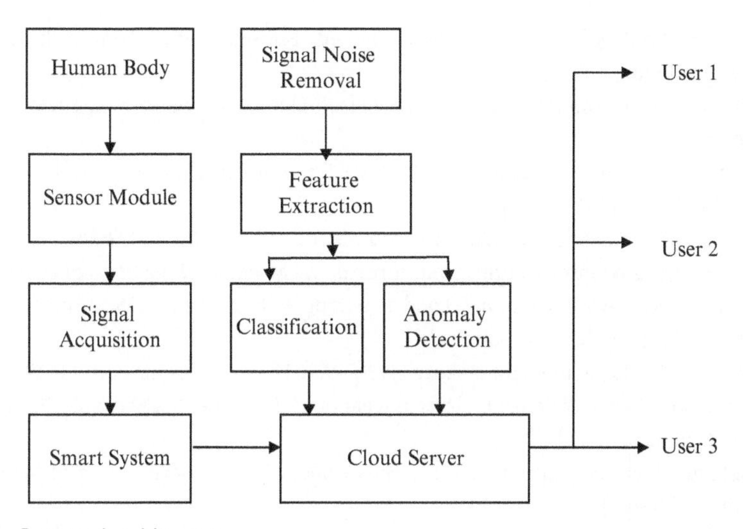

Figure 8.7 Proposed architecture.

The model can be utilized in appropriated mechanical applications for the following purposes:

- Information is prepared in disseminated design for applications in which information is divided over a far-reaching foundation restricting total information and handling on brought-together framework.
- It permits simple mix with area-explicit and conceivable stately component mappings from progressively particular artificial intelligence (AI) calculations.
- The calculation assists with demonstrating diverse mechanical large information use cases with constrained area information and low computational multifaceted nature.
- The model scales directly with the quantity of tests and period of time arrangement.

8.4 Conclusion

This chapter discusses the EHR and the various sensors used in healthcare systems. The use of sensor data and its parameters is elaborated. The various feature extraction techniques are discussed, and case studies were provided for better understanding feature extraction through sensor data.

References

[1] H. Hellman, *Beyond Your Senses*, Lodestar Books, New York, 1997. [Google Scholar].

[2] G. Buerk, *Biosensors: Theory and Applications*, Technomic, Lancaster, PA, 1993. [Google Scholar].

[3] J. Fraden, *Handbook of Modern Sensors*, 2nd ed., Springer Verlag, New York, 1996. [Google Scholar].
[4] R. Frank, *Understanding Smart Sensors*, Artech House, Norwood, MA, 1996. [Google Scholar].
[5] H. Joseph, B. Swafford, and S. Terry, "MEMS in the medical world," *Sensors*, pp. 47–51, April 1997. [Google Scholar].
[6] C. Wiebe, "Keeping close watch," *American Medical News*, June 21, 1999.
[7] Food and Drug Administration, *Summaries of the Meetings of the Clinical Chemistry and Clinical Toxicology Devices Panel*, Food and Drug Administration, Washington, DC, February 26, 1999. [Google Scholar].
[8] A. Pollack, "Stalking the wild glucose level," *New York Times*, February 11, 1999.
[9] K. Robinson, "Sensors detect biological weapons," *Photonics Spectra*, vol. 43, January 1999. [Google Scholar].
[10] A. Solovy, "Techno-treatment and the body eclectic," *Hospitals and Health Networks*, July 5, 1998. [PubMed].
[11] Anon, "A taste of the future: The electronic tongue," *CNN Interactive*, January 28, 1999.
[12] H.P. Zenner and H. Leysieffer, "Totally implantable hearing device for sensorineural hearing loss," *Lancet*, vol. 352, p. 1751, 1998. [PubMed] [Google Scholar].
[13] G. Allgood, W.W. Manges, and S.F. Smith, "It's time for sensors to go wireless," *Sensors*, pp. 70–80, May 1999. [Google Scholar].
[14] B. Pitt, et al., "Aggressive lipid-lowering therapy compared with angioplasty in stable coronary artery disease," *N. Engl. J. Med.*, vol. 341, pp. 70–76, 1999. [PubMed] [Google Scholar].
[15] E. Regis, *Nano: The Emerging Science of Nanotechnology*, Little, Brown, New York, 1995. [Google Scholar].
[16] V. Pali, S. Goswami, and L.P. Bhaiya, "An extensive survey on feature extraction techniques for facial image processing," In *2014 International Conference on Computational Intelligence and Communication Networks*, 2014. doi:10.1109/cicn.2014.43.
[17] Q. Ni, A.B. García Hernando, and I.P. De la Cruz, "The elderly's independent living in smart homes: A characterization of activities and sensing infrastructure survey to facilitate services development," *Sensors*, vol. 15, pp. 11312–11362, 2015.

9

Object Detection in Satellite Images Using Modified Pyramid Scene Parsing Networks

Akhilesh Vikas Kakade[1], S Rajkumar[2] (Corresponding Author), K Suganthi[3], and L Ramanathan[4]

[1] *SAP Labs, Bangalore*
[2,3,4] *Vellore Institute of Technology, Vellore*

9.1 Introduction

The phenomenal growth in the variety, the improved accessibility, and the global availability of satellite imagery have resulted in dramatic improvements in our understanding of planet Earth. Such an understanding is required in situations ranging from emergency operations of activating resources during disasters to routine processes like observing the impacts of global warming. However, there is still a great limitation to these developments. The limitation is the assumption that detecting features or objects of interest in satellite images can be easily done either manually or with partial help of computers, that is, semi-automatically. On the one hand, this assumption puts a tremendous burden on the experts who are responsible for detecting and identifying such objects of interest. On the other hand, there have been spectacular improvements in processing capacities of the processing units and great advancements in computer vision with help of machine learning technologies such as deep learning through deep neural networks. It is then natural to think about utilizing the hardware and logarithmic advancements in automating important or significant objects in satellite images. This identification, if automatic, accurate, and quick, can be very helpful in several applications like creating and updating maps for land use and landholding information, monitoring environmental indicators, improving urban planning, and responding disaster situations.

This chapter is inspired by the Kaggle competition "DSTL Satellite Imagery Feature Detection," announced and conducted more than 2 years ago. This chapter aims at developing a deep neural network using PSPNet architecture with modifications for detecting specified objects in satellite images provided to the Kaggle competitors. The dataset is not too large, and it is therefore considered manageable for supervised machine learning algorithms that are appropriate for problems of this nature. This chapter consists of the following major steps:

- Adaptation of convolutional neural networks (CNNs) to multispectral image data and evaluation of data fusion strategies for semantic segmentation on satellite images.
- Introduction of a combined training objective for defining the desired output for the purpose of image segmentation.

Sensor Data Analysis and Management: The Role of Deep Learning, First Edition. Edited by A. Suresh, R. Udendhran, and M.S. Irfan Ahmed
© 2021 John Wiley & Sons, Ltd. Published 2021 by John Wiley & Sons, Ltd.

9.2 Problem Statement

Satellite images contain a huge amount of data, both visible and invisible. A variety of methods have been developed for extracting information from satellite images due to different applications like environmental monitoring, urban and rural development planning, management of natural resources, and many more. One of the recent trends is to extract information from image data for the purpose of security in the form of detecting and tracking vehicles and identifying illegal constructions, water bodies, roads and other tracks, and so on. The Kaggle competition required the participants to identify objects of the following types:

1. Buildings
2. Miscellaneous artificial structures
3. Highways
4. Pathways (cart tracks)
5. Trees (stand-alone trees, group of trees, etc.)
6. Crop (cereal crops such as wheat, maize; row crops such as potatoes and turnips; contour ploughing; and cropland)
7. Canals
8. Stagnant water
9. Huge vehicles (truck, bus, trailer, etc.)
10. Small vehicles (car, van, motorcycle, etc.)

After reviewing the literature, it was decided to generate feature masks for different objects. The reason for generating masks was to achieve a semantic segmentation by only detecting categories of different objects without identifying the objects individually.

9.3 Data Overview

Defense Science and Technology Laboratory (DSTL) has provided 1 km × 1 km satellite images in panchromatic, three-band (RBG), and two–eight–four-band formats. The total number of images provided by DSTL is 450. The number of training images is 25; 32 images are utilized in a test set, and the remaining images in the validation set. Every image is available in the three versions. Table 9.1 gives more information on the three versions.

The two 8-band channels have to be resized and aligned to match the 3-band channels. All channels are then concatenated to form a single 20-channel input image for processing.

Table 9.1 Three versions of a satellite image.

Type	Wavebands	Pixel resolution	No. of channels	Size
Grayscale	Panchromatic	0.31 m	1	3348 × 3392
3-band	RGB	0.31 m	3	3348 × 3392
8-band	Multispectral	1.24 m	8	837 × 848
8-band	Short-wave infrared	7.5 m	8	134 × 136

The reason for utilizing all the 20 inputs is that every channel covers a unique range of the spectrum and hence records unique features that other channels are not capable of observing. Table 9.2 gives specifications and unique features of different channels of the WorldView-3 satellite.

The spectral resolution of these images is also higher due to these having 11-bit and 14-bit depth for every pixel instead of the traditional 8-bit depth of earlier satellites. It is also important to note that images in different channels are captured at different time points.

Table 9.2 Specifications and unique features of different channels.

Band/channel	Range	Unique features
Panchromatic	400–800 nm	Sharp image due to 31 cm resolution sacrifices spectral resolution for 1.24 m spatial resolution
Multispectral bands		
Coastal	400–450 nm	Absorbed by chlorophyll, least absorbed by water, and substantially influenced by atmospheric scattering
Blue	450–510 nm	Absorbed by chlorophyll, produces better penetration of water, and also withstands against atmospheric scattering
Green	510–580 nm	Can focus more precisely on healthy vegetation, ideal for calculating plant vigor
Yellow	585–625 nm	Plays a pivotal role in feature classification, spots the "yellowness" of certain vegetation on farmland as well as in water bodies
Red	630–690 nm	The absorption of red light is concentrated more by chlorophyll, important for differentiating the vegetation, and useful in categorizing bare soils, roads, and geological features
Red edge	705–745 nm	Centered at high reflectivity of vegetation, and very useful in calculating plant health and classifying the vegetation
Nitrite Reductase 1 (NiR1)	770–895 nm	Very productive in measuring the moisture and biomass in plants, effectively segregates water bodies from vegetation, recognizes types of vegetation, and distinguishes soil types
NiR2	860–1040 nm	Less affected by atmospheric influence, and provides wide vegetation analysis and biomass studies
SWIR-1	1195–1225 nm	Sensitive to moisture in soil and vegetation, and useful in separating wet from dry earth. Can see through smoke. Makes it easy to obtain spectral signatures of several minerals
SWIR-2	1550–1590 nm	
SWIR-3	1640–1680 nm	
SWIR-4	1710–1750 nm	
SWIR-5	2145–2185 nm	Helps monitor blue-green algae and turbid water. Ideal for understanding crop water stress. Hydrous minerals appear darker
SWIR-6	2185–2225 nm	
SWIR-7	2235–2285 nm	
SWIR-8	2295–2365 nm	

9.4 Literature Survey

In semantic segmentation as presented in Figure 9.1, we take an image and divide it into meaningful parts. Each part is examined at the pixel level and classified into a predefined class. Most of the times, deep learning techniques are used for such classification. One such deep learning technique used in semantic segmentation is CNN. CNN is a supervised classification method that can learn the important features of an image in an end-to-end manner. It can also learn optimum features very quickly and does not underperform even if the underlying image has minor variations.

Khan [1] proposed a target detection system for satellite imagery that uses EdgeBoxes and CNN for classifying target and nontarget objects in a scene. The edge information of targets in satellite imagery contains very prominent and concise attributes. EdgeBoxes use the edge information to filter the set of target proposals. The prediction was limited to two-class objects whether it is an artifact or nonartifact. The proposed model cannot be used for multiclass object detection.

There have been advancements in image segmentation in DeepLabv3, which implements a ResNet model using dilated/atrous convolutions [2]. We have chosen architectures such as modified PSPNet and U-net over DeepLabv3, as they have familiar implementations using concepts learned in object class; use comparably less parameters; and trains faster than deep convolution neural networks (DCNN) models such as DeepLab and other pixel-level classifiers.

In real-time applications, detecting an object is critical. A variant of CNN—the faster R-CNN [3]—can be used in real-time applications to detect objects quickly. Due to region of interest (RoI) pooling, the estimation of convolution layers between proposals is shared. The system is trained completely. Since another model generates the region proposals, the enhancement in speed is not high in the faster R-CNN.

9.5 Methodology

9.5.1 Preprocessing Steps

The preprocessing of images involves the following four steps:

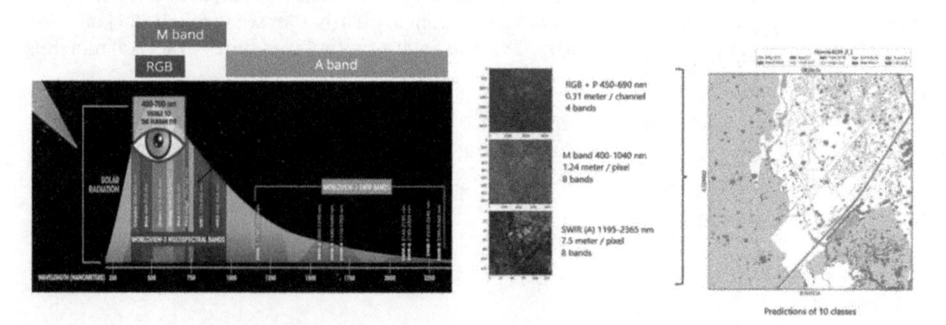

Figure 9.1 The spectrum converge of the four versions of images and their mutual connection.

Step 1: The four versions of an image—namely, the panchromatic (1 band), RGB (3 bands), multispectral (8 bands), and short-wave infrared (SWIR) (8 bands)—go for input and are synchronized before being concatenated for further analysis, as presented in Figure 9.2.
Step 2: Images in training data are subjected to the scale percentile process to make them comparable to other images.
Step 3: The concatenated 20-channel image is converted to a multipolygon WKT format for creating masks for different objects for easy detection.
Step 4: Patches of a specified size are selected from images for training the CNN.

The last step is necessary because the original images are too large for training the CNN. This study has used 224 as the input size for CNN. This size has been arrived at after carrying out experiments.

9.5.1.1 Scale Percentile Processing

The 3-band (RGB) images have mostly 8-bit spectral resolution. That is, each pixel has 256 levels, from 0 to 255. For example, a completely red pixel has the spectral value (255, 0, 0), a completely green pixel (0, 255, 0), and a completely blue pixel (0, 0, 255), while a white pixel has the spectral value (255, 255, 255), and the perfectly black pixel has the spectral value (0, 0, 0). More recent satellite sensors have 11-bit or 14-bit images and hence have a larger range of spectral pixel values. Of course, these images can store more information, but they also require larger storage spaces. This feature may also have compatibility issues with software.

The images used in this study are converted from 14-bit to 8-bit spectral resolution. Even through software tools like NumPy are available for this conversion, this study has made use of the GDAL library. The scale percentile process normalizes image luminance and resizes the input image to a square image that has side length of 112. The resizing does not affect the aspect ratio and preserves it. This process specifies a percentile range of 1–99, and pixels having spectral values outside this range (i.e., below the 1st and above the 99th percentiles) are removed, since they are declared to be outliers. The cleaned image is then rescaled to 8-bit spectral resolution.

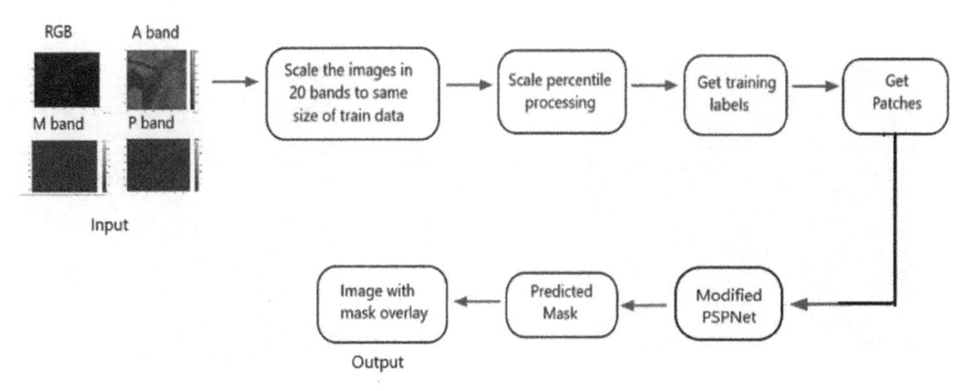

Figure 9.2 Proposed pipeline.

Let X_{in} be an 8-bit input band whose value ranges from 0 to 255, say a = 0 and b = 1, and let X_{out} be the output band. For scale percentile processing, we choose lower percentile and higher percentile as c and d to be 1st and 99th percentile, respectively. The scale percentile processing can be represented using the following function:

$$X_{out} = (X_{in} - c)\frac{(b-a)}{(d-c)} + a$$

We clip the values to minimum and maximum, such as

$$X_{out}\,[X_{out} < a] = a \text{ and } X_{in}[X_{in} > b] = b$$

9.5.1.2 Patches and Input

The literature on satellite imagery mentions the panchromatic band as p-band, the 8-channel multispectral band as M-band, and the 8-channel short-wave infrared band as A-band. The four bands do not have the same resolution and have therefore to be resized for spatial synchronization. It is also found that the frequency distributions of the 10 objects to be detected are skewed in the images in the training set. The number of images in the training set is 25 and is not enough for training. Further, the size of every image is too large for processing, as presented in Figure 9.3. Considering all issues related to images, every image is divided into square parts of size 112 × 112, and these parts are called *patches*. Object detection is then carried out on these patches rather than on original images. This has allowed the deep neural network to train properly due to large training data while enhancing the processing speed due to the reduced size of every individual input data element.

9.5.2 Deep Neural Network Architectures

Object detection in the given satellite images is the objective of this study, and training data has been provided to develop the classification rules. However, there is no evidence in the literature that one particular deep neural network architecture is optimal. This study has deployed four different architectures, such as multispectral U-net architecture, inverted

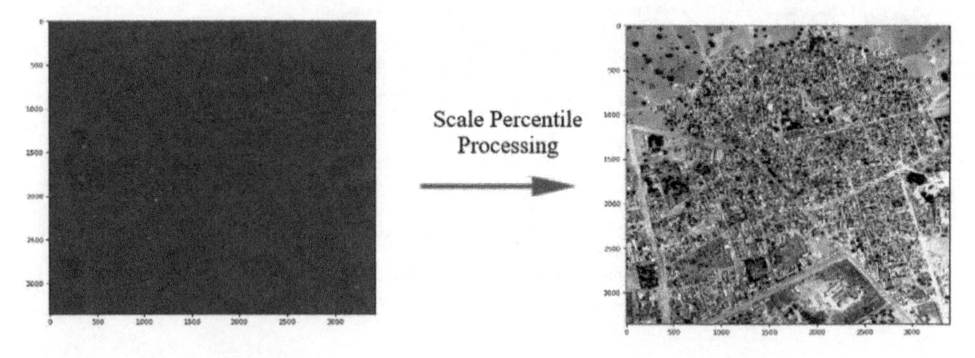

Figure 9.3 Preprocessing step.

pyramid, modified PSPNet, and XGBoost algorithm, so that the best can be identified at the end of the study. These four architectures are briefly described here.

9.5.2.1 Multispectral U-net Architecture Model

The literature indicates that most of the image analysis and classification problems are solved with deep CNN. The U-net architecture [4] gives a fully connected CNN. It was developed by a German research team at the University of Freiburg for biomedical image segmentation. It has shown good performance at image segmentation for the problem of nerve detection in ultrasound images [5].

The U-net architecture has contractive and expansive paths. The architecture of contractive paths is usually a convolution neural network. Batch normalization has been used in this study for accelerating convergence during training. Also, the primary activation function is exponential instead of linear (exponential linear unit [ELU]) [6]. The following figure shows the multispectral U-net architecture used in this study:

The following hyperparameters are used during training the U-net model: batch size is set to 16, primary activation function used is ELU [6], learning rate is set to 0.00001, optimizer used is Adam, and loss used is "binary_crossentropy," as presented in Figure 9.4.

9.5.2.2 Inverted Pyramid Model

The inverted pyramid architecture is an experimental model, and it has been used in this study as the second option. The image decreases in size as it passes through the network. Every path of the network uses different parts of the image, while the output size is fixed. Dilated convolution is used in the early part of the network for decreasing the image size while retaining a larger receptive field for output neurons. Dilated convolutions are particularly popular in the field of real-time segmentation. The Cropping2D layer is used for 2D input layer, and it crops along spatial dimensions, that is, height and width. Different paths are combined later in the network. Dropout is used at final layers for added regularization, as no max pooling is used. All Conv2D layers are represented as conv->batch norm->elu, but the extra layers are suppressed to make viewing slightly easier. This architecture is still in the experimental stage, and is shown in Figure 9.5.

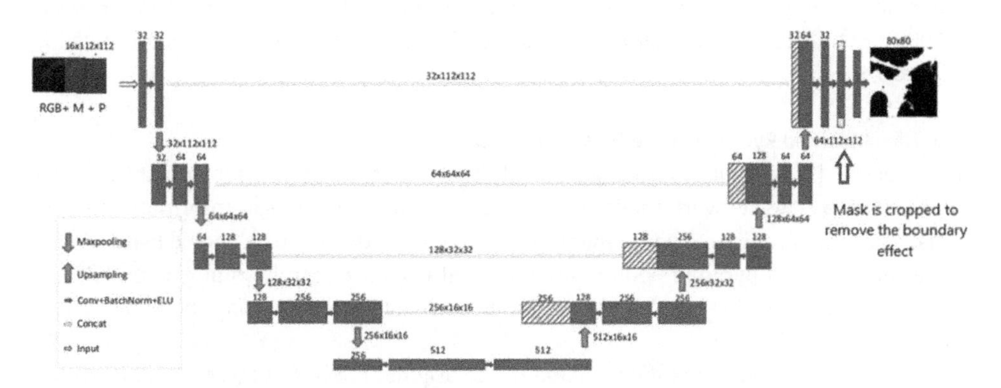

Figure 9.4 Multispectral U-net architecture.

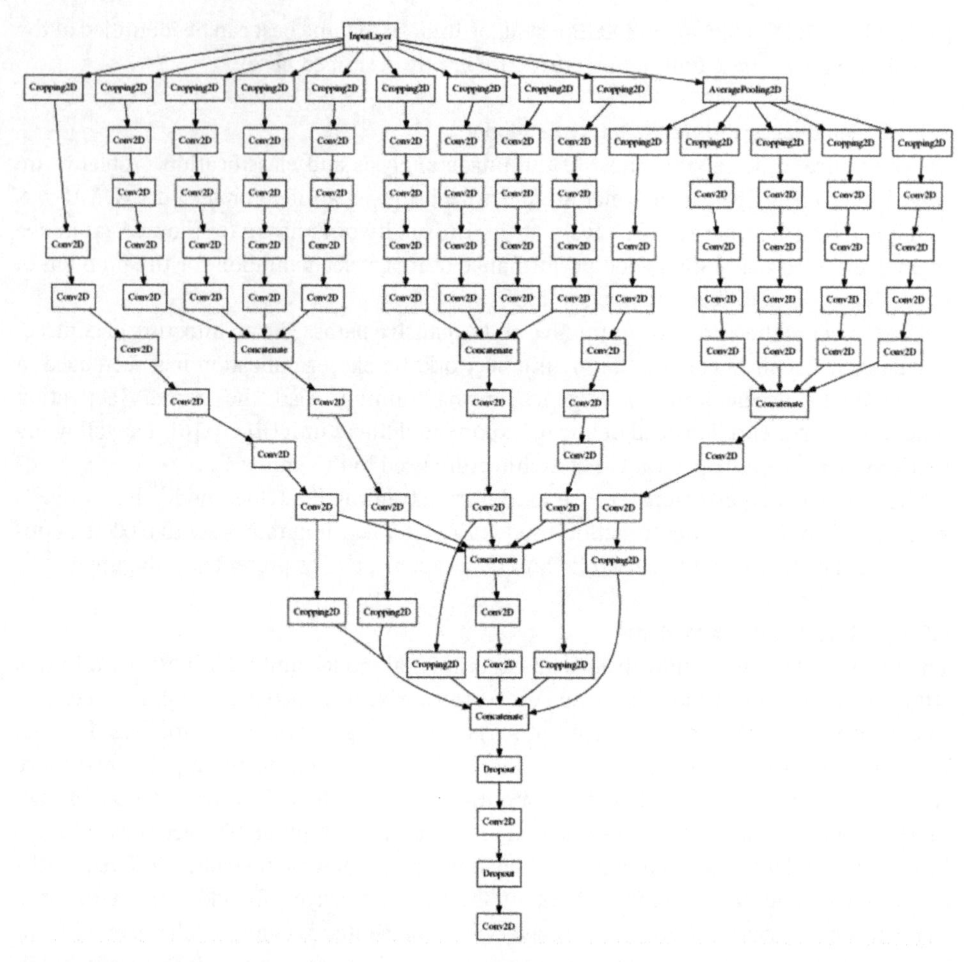

Figure 9.5 Inverted pyramid.

The following hyperparameters are used during training the inverted pyramid model: batch size is set to 16, primary activation function used is ELU [6], learning rate is set to 0.00001, optimizer used is Adam, and loss used is "binary_crossentropy," as presented in Figure 9.5.

9.5.2.3 Modified Pyramid Scene Parsing Network

Zhao et al. [7] introduced this advanced deep learning model for semantic image segmentation. This network model focuses on utilizing geographical information at various scales when compared with other CNN models. In order to create pooled feature maps at various stages, a pyramid pooling module on the feature map is employed by PSPNet, which is generated by ResNet [8], wherein these features are combined for further analysis.

This study has used a modified version of the original PSPNet. As part of the modification, an encoder–decoder is added before the CNN for converting the 20-channel input image into a 3-channel image for input to the CNN. In order to decrease the specified depth level of the pyramid, the four levels of the pyramid pooling module feature with different bin sizes, and these are attached to the feature map generated by ResNet. A 1×1 convolutional layer is added to each pyramid level. The total of channels of the four levels is equated to the dimension of original feature map. The resulting five feature maps are concatenated to get one feature map that can be used for the final stage of analysis, namely, object detection.

In Figure 9.6, given an input image (a), we first pass the input layer of 20-channel input image data to (b) encoder–decoder to convert it into an output layer of 3 channels. This helps to preserve the maximum features during prediction and to avoid the need for heavy computations to be carried out for the huge 20-channel input layer. Then, we pass the 3-channel layer as an input to CNN, that is, ResNet-101 gets the (c) feature map of the last convolutional layer. Then, (d) a pyramid parsing module is applied to harvest different sub-region representations, followed by upsampling and concatenation layers to form the final feature representation, which carries both local and global context information in (e). Finally, the representation is fed into a convolution layer to get the final per-pixel prediction.

The following hyperparameters are used in training the PSPNet model: batch size is set to 16, primary activation function used is rectified linear unit (ReLu) [9], learning rate is set to 0.00001, optimizer used is Adam, and loss used is "binary_crossentropy."

9.5.2.4 XGB Classifier

XGBoost decision tree [10, 11] is a more traditional machine learning algorithm that uses aggregate image features for training. The algorithm is used as the fourth model in this study for comparing its performance with that of the other three methods.

The data in training images consist of locations of objects that are to be detected. Every image is divided in several grids, and the best grid is chosen with a view to improve the classification score. Jaccard index is used for these comparisons. A feature vector is formed for every grid, where the included features are mean, variance, skewness, and kurtosis. These four features are extracted for every channel. Since there are 10 objects of interest, 10 trees are created to determine if a label should be attached to an image. The trees have had a maximum depth of 5 with 100 estimators per label.

After carrying out all analyses, it is found that training one separate CNN for every object class achieves much higher accuracy than training a single CNN for all the 10 object classes.

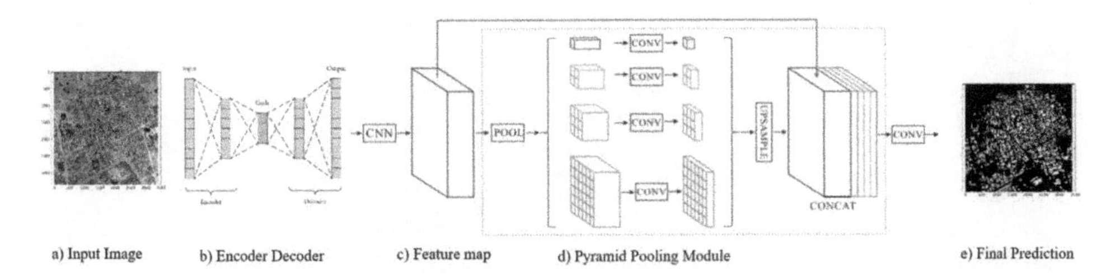

Figure 9.6 Overview of our proposed PSPNet.

9.6 Evaluation Metrics

For evaluating our classification results, we are using two metrics as follows:

(i) Pixel accuracy

Pixel accuracy assesses our outcomes by simply identifying the number of pixels which were effectively classified in an image. Pixel accuracy is ordinarily revealed per class and for overall classes.

When per-class pixel is taken into consideration, we are basically evaluating the binary mask, where true positive outcomes pixels of an image that are correctly classified for the given class when it is compared with ground truth mask, and true negative outcomes pixels of an image that are correctly classified but do not belong to the given class.

$$\text{Pixel accuracy} = \frac{\text{TP+TN}}{\text{TP+TN+FP+FN}}$$

The evaluation metric used can sometimes give wrong results when the class object present is small in the image. Model efficiency is tested well when there is negative case, that is, when class is not present in an image.

(ii) Jaccard index

The Jaccard index measures the similarity among a limited number of sets, which is also known as *intersection over union* (IOU). IOU is a statistic measure used for comparing the similarity and diversity measure between two sets—A and B—which can be defined as:

$$J(A, B) = \frac{|A \cap B|}{|A \cup B|} = \frac{|A \cap B|}{|A| + |B| - |A \cap B}$$

$$0 \leq J\,(A, B) \leq 1$$

To analyze the performance of the proposed algorithm of all the labels, we must calculate the Jaccard index of each label and take the average as presented in Figure 9.7, which is

$$\text{Score} = \sum_{i=1}^{10} \text{Jaccard i}$$

Overall, the problem can be viewed as a classic supervised image classification and object recognition problem with multispectral input image channels and score function with Jaccard index.

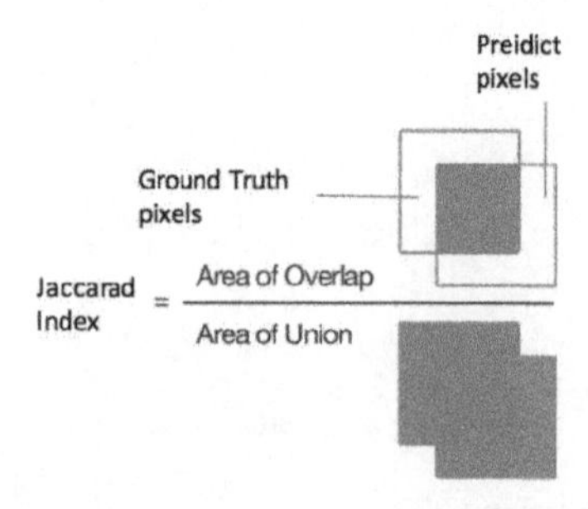

Figure 9.7 Jaccard coefficient.

9.7 Results and Analysis

(i) Multispectral U-net architecture:
(ii) Inverted pyramid architecture:
(iii) XGBoost classifier:
(iv) Modified pyramid scene parsing network:
(v) Comparison of predicted masks for all architectures in class 7:

Table 9.3 shows the results per class, which indicates that the modified PSPNet architecture outperforms the other approaches with respect to pixel accuracy as well as mean IOU. Apart from the addition of an encoder–decoder, the size of the kernel is also changed in the convolution layer after the process of pyramid pooling from original 3×3 to 1×1. This saves on computational performance while no major difference is seen in accuracy measure as presented in Figures 9.8–9.12. We have also modified the hyperparameters, namely, the batch size is set to 16 and the base learning rate is set to 0.00001; ReLu [9] is used as primary activation function instead of ELU [6], which proved beneficial for the overall training and validation scores.

Table 9.4 shows that our approach gives better results for overall classes as compared to Vladimir Iglovikov's approach.

Table 9.3 Results for each class in terms of pixel accuracy and mean IOU on test data.

Architectures	**Multispectral U-Net**		**Inverted pyramid**		**Modified PSPNet**		**XGBoost**	
Classes	Pixel accuracy	Mean IOU	Pixel accuracy	Mean IOU	Pixel accuracy	Mean IOU	Pixel accuracy	Mean IOU
Buildings	0.9418	0.6484	0.8957	0.6597	0.8317	0.7500	0.5296	0.4150
Structures	0.9699	0.5046	0.2095	0.0175	0.9692	0.4118	0.3434	0.0140
Roads	0.9560	0.7639	0.7190	0.5450	0.8392	0.7694	0.5276	0.3440
Tracks	0.9528	0.5305	0.5580	0.1790	0.9311	0.5451	0.1657	0.0380
Trees	0.9644	0.6713	0.9095	0.6008	0.8644	0.7713	0.6447	0.5109
Crops	0.8734	0.8260	0.9176	0.8865	0.8566	0.8033	0.6919	0.6485
Waterways	0.9610	0.81110	0.8521	0.7709	0.9205	0.8913	0.8016	0.5739
Standing water	0.9238	0.7087	0.8688	0.5268	0.9352	0.7123	0.6830	0.4769
Large vehicles	0.6550	0.4000	0.3807	0.2248	0.9307	0.6069	0.24062	0.0265
Small vehicles	0.6958	0.4978	0.3217	0.1056	0.9158	0.5940	0.2666	0.0038
	Average pixel accuracy	Average IOU	Average pixel accuracy	Average IOU	Average pixel accuracy	Average IOU	Average pixel accuracy	Average IOU
All classes	0.8893	0.6362	0.6632	0.4516	0.8994	0.6855	0.4894	0.3051

Table 9.4 Result comparison of multispectral U-net for different classes of DSTL datasets in terms of intersection over union with our approach and Vladimir Iglovikov's [12] approach.

Class	Test data (Vladimir Iglovikov's approach)	Test data (our approach)
Buildings	0.7453	0.6484
Structures	0.1905	0.5046
Roads	0.8005	0.7639
Tracks	0.3281	0.5305
Trees	0.5018	0.6713
Crops	0.8251	0.8260
Waterways	0.9697	0.81110
Standing water	0.6081	0.7087
Large vehicles	0.2964	0.4000
Small vehicles	0.0186	0.4978
	Average IOU	Average IOU
All classes	0.52841	0.63623

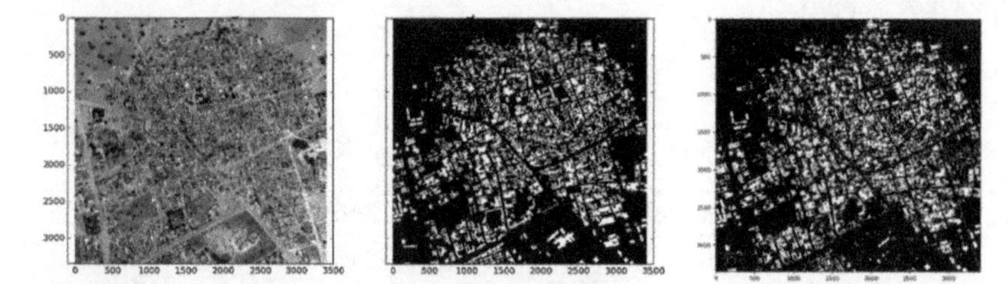

Figure 9.8 Predicted mask for class 1.

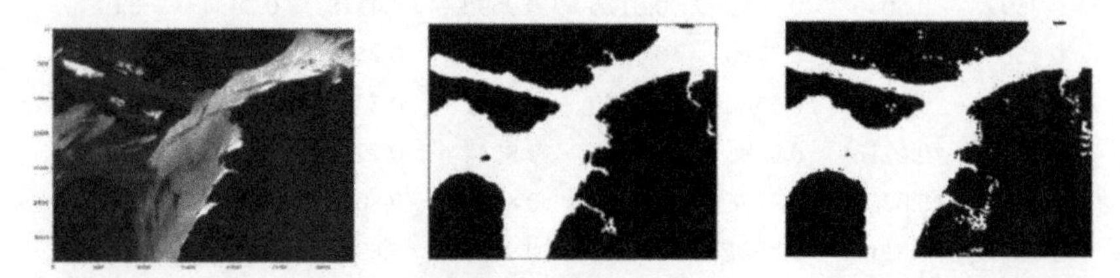

Figure 9.9 Predicted mask for class 7.

Figure 9.10 Predicted output for class 5.

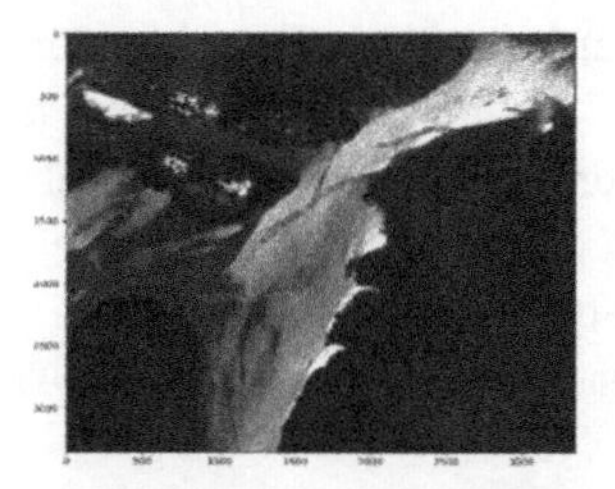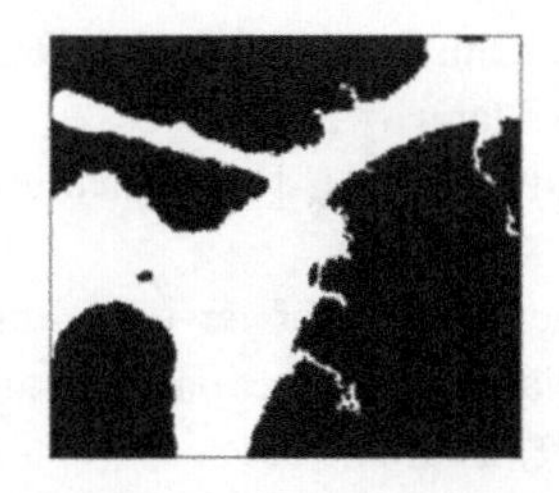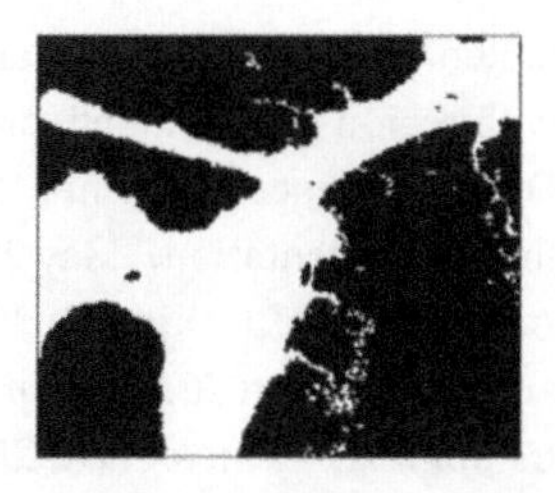

Figure 9.11 Predicted output for class 7.

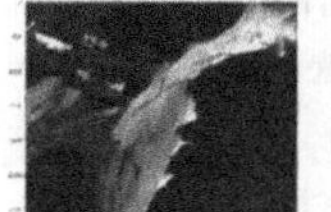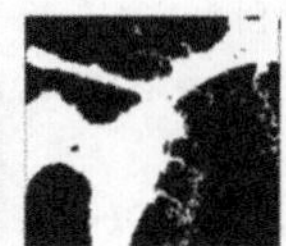

Figure 9.12 The abovementioned comparison of predicted masks shows output performances of all architectures on an image containing class 7, that is, a waterway; PSPNet gives good results in terms of intersection over union when compared with ground truth mask.

9.7 Conclusion

In this research, we used real-world data to deal with problems and tried to solve it with deep learning methods. The dataset is cluttered, and we spent most of the time dealing with data preprocessing. Our method of introducing an encoder–decoder to PSPNet improves overall computation performance and accuracy; additional changes like 1×1 convolution kernel instead of 3×3 convolution kernel are used. ReLu [9] is deployed as primary activation function instead of ELU [6], which also helped in overall computation and accuracy. The accuracy of training each class separately with 1 CNN is much higher than training all 10 classes at a time with 1 CNN. U-net architecture already has a heavyweight decoder, since it has the same number of parameters as its encoder. The modified PSPNet module seems to have a better encoder–decoder, and U-Net would need additional decoder capacity.

The proposed approach has successfully handled the abovementioned problems, which significantly improves the quality of the final models. Our work incorporates several steps such as the adaptation of fully convolutional networks to multispectral satellite images and the evaluation of several data fusion strategies on semantic segmentation tasks of satellite images with a combined training objective.

References

[1] M.J. Khan, "Automatic target detection in satellite images using deep learning," Article in *J. Space Technol.*, July 2017.

[2] L. Chen, G. Papandreou, F. Schroff, and H. Adam, "Rethinking atrous convolution for semantic image segmentation," December 2017

[3] S. Ren, K. He, R. Girshick, and J. Sun, “Faster R-CNN: Towards real-time object detection with region proposal networks,” January 2016.
[4] O. Ronneberger, P. Fischer, and T. Brox, “U-Net: Convolutional networks for biomedical image segmentation,” May 2015.
[5] A. Kakade and J. Dumbali, “Identification of nerve in ultrasound images using U-net architecture,” In *2018 International Conference on Communication information and Computing Technology (ICCICT)*, February 2018.
[6] D.-A. Clevert, T. Unterthiner, and S. Hochreiter, “Fast and accurate deep network learning by exponential linear units (ELUs),” February 2016.
[7] H. Zhao, J. Shi, X. Qi, X. Wang, and J. Jia, “Pyramid scene parsing network,” April 2017.
[8] K. He, X. Zhang, S. Ren, and J. Sun, “Deep residual learning for image recognition,” December 2015.
[9] A.F.M. Agarap, “Deep learning using rectified linear units (ReLU),” March 2018
[10] T. Chen and C. Guestrin, “XGBoost: A scalable tree boosting system,” June 2016.
[11] A. He, J. He, and R. Kim, “An ensemble-based approach for classification of high-resolution satellite imagery of the Amazon basin,” July 2017.
[12] V. Iglovikov, S. Mushinskiy, and V. Osin, “Satellite imagery feature detection using deep convolutional neural network: A Kaggle competition.”

10

Coronary Illness Prediction Using the AdaBoost Algorithm

G. Deivendran, S. Vishal Balaji, B. Paramasivan, S. Vimal (Corresponding Author)

National Engineering College, TamilNadu, India

10.1 Introduction

Diseases may be defined as a range of conditions that affect the normal functioning of the body. Diseases can be divided into four main categories, which include genetic disease, infectious disease, deficiency disease, and physiological disease. Among these categories, heart disease, also known as *cardiac disease*, falls under the category physiological disease. There are many risk factors that lead to the malfunctioning of the heart, which include smoking, lifestyle, blood pressure, etc. [1] Heart diseases include heart failure, heart attack, arrhythmia, artery diseases, rhythm problems, etc. Heart disease has become one of the most serious health issues worldwide. According to a recent survey by the World Health Organization, 35% of global deaths in a year are due to cardiac illness. Out of these, 85% are due to heart attacks and strokes. India occupies 56th position, with 14% of total deaths owing to heart diseases.

Clinical diagnosis of heart diseases includes examination of the patient with symptoms and the available test results by a medical practitioner. The results of blood samples, electrocardiograph, echo, treadmill, and angiography are used to diagnose the diseases [2]. The accuracy of manual diagnosis is subject to the knowledge and experience of the cardiologist and the accuracy of the test results, which may sometimes lead to wrong diagnosis, which may be fatal [3]. Individuals with cardiovascular disease or who are at high cardiovascular hazards need early identification. Hence, there is a need for automation of heart disease prediction, which must also be accurate.

ML is one of the emerging technologies used in various fields such as video surveillance, healthcare, social media analysis, spam detection, etc. ML is an application of artificial intelligence (AI) [4], and is used to execute AI. AI is a system for data examination that mechanizes an indicative model structure. Today, AI streamlines managerial procedures in clinics, guides and treats irresistible ailments, and customizes medications [5]. The major difference between ML and data mining is that ML focuses on prediction, while data mining involves the discovery of unknown properties in available data [6]. ML uses data mining as a preprocessing step to improve accuracy.

Sensor Data Analysis and Management: The Role of Deep Learning, First Edition. Edited by A. Suresh, R. Udendhran, and M.S. Irfan Ahmed.
© 2021 John Wiley & Sons, Ltd. Published 2021 by John Wiley & Sons, Ltd.

The application of ML in the healthcare industry is enormous. It is used in disease diagnosis, drug discovery, health record maintenance, medical image diagnosis, etc. There are various ML algorithms such as naïve Bayes (NB), random forest (RF), k-nearest neighbor, decision tree (DT), support vector machine (SVM) [7], and many others that are used in the prediction of heart disease [8]. Various ML algorithms provide different accuracies. Neural networks prove to be efficient in heart disease prediction. Neural networks can also be used with other ML algorithms for improved accuracy [9, 10].

In this chapter, we use a classification algorithm known as the AdaBoost algorithm. The fundamental goal of this examination is to improve the accuracy of coronary illness forecast. This algorithm increases the weights for the misclassified elements, thereby decreasing the error rate which, in turn, increases the efficiency of the coronary illness prediction.

The chapter is organized as follows: Discussions related to heart disease and existing methods are presented in Section 10.2. Section 10.3 deals with implementation details, which include preprocessing, feature selection, classification, performance measures, and results. Section 10.4 concludes with the current work and includes notes on future enhancement. Section 10.5 contains the references for this chapter.

10.2 Literature Survey

Mahboob et al. [8] worked on ML algorithms such as hidden Markov model, support vector machine, feature selection, computational intelligent classifier, predicting system, data mining techniques, and genetic algorithm, and proposed an ensemble model. The ensemble model proved to be more efficient when compared to the algorithms like k-nearest neighbor, artificial neural network (ANN), and SVM. The proposed model provides an accuracy, root mean square error (RMSE), receiver operating characteristics curve (ROC), and precision of 94.21%, 0.981, 0.2568, and 0.953, respectively.

Shao et al. [10] proposed a hybrid model that includes ML algorithms such as logistic regression (LR), multivariate adaptive regression splines (MARS), ANNs, and rough set (RS) techniques. It consists of two stages. In the first stage, algorithms such as LR, MARS, and RS techniques are used to reduce the set of explanatory variables. In the second stage, the remaining variables are fed as input to the ANN. The result of the proposed hybrid model proved to be more efficient than the single-stage ANN method.

Amin et al. [11] worked on the identification of significant feature selection, as the features have a great impact on the result obtained, that is, accuracy. They developed a model using the different combinations of features and classification techniques such as decision trees, naïve Bayes, logistic regression, support vector machine, and hybrid models (which includes naïve Bayes and logistic regression). The result obtained identified nine significant features and the best performing algorithm—VOTE, which achieved an accuracy of 87.4%.

Sabahi [1] proposed the bimodal fuzzy analytic hierarchy process (BFAHP). In this model, the reciprocal comparison matrix is constructed by using fuzzy validity and fuzzy probabilities, which are calculated by aggregating the validity of relevant risk and Bayesian formulation, respectively. The decision of being affected and not affected is made using the ranking of high and low risk. The model was applied to an Iranian dataset of 152 patients and obtained an accuracy of 85%. The results also confirm that adding validity in a fuzzy manner increases the confidence level.

Vivekanandan and Iyengar [12] proposed the differential evolution algorithm, which is used for the selection of critical features from the dataset, since feature selection is an important phase in data preprocessing. The result proved the efficiency of the algorithm with 83% accuracy. The time factor is also analyzed, which also proved efficient.

Wagh and Paygude [4] proposed that the application of AI in the prediction of heart disease provides improved results. They used the neuro-fuzzy model for prediction. It also uses the global optimization advantage of the genetic algorithm for the initialization of neural network adaptive capabilities and fuzzy logic reasoning approach for effective prediction. A combination of neuro fuzzy systems (NFS) and genetic algorithm (GA) provides an accuracy of 90%, while the NFS alone provides 82% accuracy.

Samuel et al. [13] proposed x^2-GNB for the effective prediction of heart disease. This feature-driven decision support system consists of two phases. In the first phase, the x^2 statistical model is used to rank the 13 attributes. The obtained x^2 test score is used to search an optimal subset of features by the forward best-first search strategy. The second phase includes the implementation of the Gaussian naïve Bayes. The proposed model is found to be 93.33% accurate when evaluated on a dataset with 297 entries.

Nahar et al. [7] investigated many computational intelligence techniques to predict cardiac disease. The expert judgment-based multi-label feature selection (MFS) was highlighted. It was also found that the well-known Cleveland dataset is imbalanced when considering binary classification. The results proved that MFS had improved efficiency. It also proved that MFS combined with the computerized feature selection process is more efficient for naïve Bayes, IBK, and SMO.

Das et al. [9] introduced a new method that includes SAS-based software for diagnosing. A neural network ensemble model is used, which works by combining the posterior probabilities, which are also known as *predicted values* from multiple predecessor models. This proposed model also proved to be effective. Three independent neural networks were used to construct the ensemble model. The results obtained include 89.01% accuracy, 80.95% sensitivity, and 95.91% specificity values.

Baccour [14] proposed a new classifying algorithm, amended fused TOPSIS-VIKOR for a classification (ATOVIC), which is a combination of multicriteria decision-making (MCDM) and VIKOR. In this proposed method, criteria are replaced by features, and alternatives are replaced by objects. Reference and test sets are employed. The proposed model is applied to the Cleveland dataset by utilizing binary classification and multiclassification. The results obtained contain improved true positive values, and it can hence be used for prediction and classification.

10.3 Implementation

The implementation consists of five stages, which include preprocessing, feature selection, classification using AdaBoost, performance analysis, and result. Among these five stages, classification is considered the most important [14]. Figure 10.1 illustrates the implementation stages.

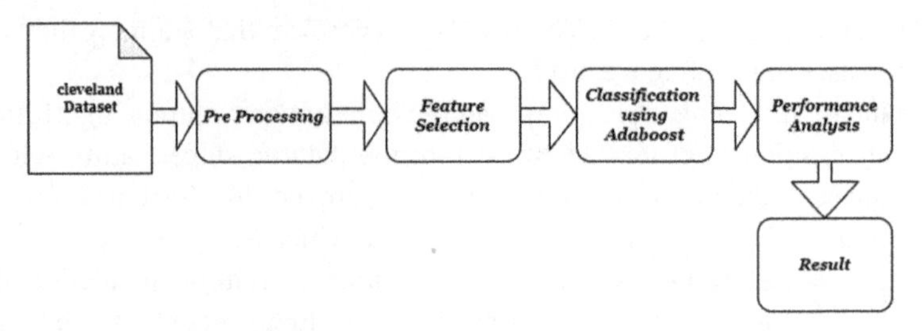

Figure 10.1 Implementation flowchart.

10.3.1 Preprocessing

Preprocessing of data is a step followed to make a dataset suitable for classification using any ML algorithm. Usually, a dataset consists of data with many missing values, which makes the dataset unfit for processing using certain ML algorithms. Hence, those missing values are to be treated or omitted during classification. The dataset consists of many attributes, each with different ranges. This range can be categorized or binarized. In our dataset, out of 303 records, 6 records contain missing values that are omitted, and the remaining 297 records are taken for classification [15]. Some attributes are binarized, some are categorized, and some are taken as the original value for classification.

10.3.2 Feature Selection

Feature selection is considered an important process because it has great impact on the performance of the model [11, 12]. In the Cleveland dataset, there are 76 attributes, from which 13 attributes are identified as important and are taken for classification [16]. Out of these 13 attributes, age and sex are clinical records, while the remaining 11 records are nonclinical records. Table 10.1 provides the description and range of various attributes used for classification.

10.3.3 Classification

10.3.3.1 Naïve Bayes

Strong and naïve independence assumptions are utilized in naïve Bayes for classification [8]. It is assumed that some features of a class are not correlated. Only a small training set is required to evaluate the mean and variance, which is required for classification. The conditional and marginal probabilities of two arbitrary events are related in this classification. It is also used frequently to compute the posterior probabilities of the given records [13]. This classification technique assumes that the presence of a particular feature of a category is irrelevant to the presence of another element:

$$P(c|x) = \frac{P(x|c)P(c)}{P(x)} \tag{10.1}$$

Table 10.1 Attribute information.

S. no.	Attribute	Description	Range
1	Age	Age of the person in years	Above 19
2	Sex	0 = female/1 = male	0,1
3	Cp	Chest pain 1 = typical angina/2 = atypical angina/ 3 = nonanginal pain/4 = asymptomatic	1,2,3,4
4	Trestbps	Resting blood pressure	Numeric value measured in mm/Hg
5	Chol	Serum cholesterol	Numeric value measured in mg/dl
6	Fbs	Blood sugar level >120 mg/dl 1 = true/0 = false	0,1
7	Restecg	Resting electrocardiographic results	0,1
8	Thalach	Maximum heart rate achieved	Numeric value
9	Exang	Exercise-induced angina 1 = yes/0 = no	0,1
10	Oldpeak	ST depression induced by exercise relative to rest	Numeric value
11	Slope	The slope of the peak exercise ST segment 1 = upsloping/2 = flat/3 = downsloping	1,2,3
12	Ca	Number of major vessels (0–3) colored by fluoroscopy	0,1,2,3
13	Thal	Status of heart illustrated through three valves 3 = normal/6 = defect/7 = reversible defect	3,6,7
14	Num	Heart disease in five valves 0 = absent/1 to 4 = present	0,1,2,3,4

$$P(c \mid X) = P(x_1 \mid c) \times P(x_2 \mid c) \times \ldots \times P(x_n \mid c) \times P(c)$$

Bayes' theorem gives a method of computing the posterior probability P(c|x), which is calculated using P(c), P(x), and P(x|c). The assumption of the naïve Bayes classifier is that the impact of the estimation of a predictor (x) on a given class (c) is independent of the estimation of other predictors. This assumption is called *class conditional independence*:

- P(c|x) is the posterior probability of class given predictor.
- P(c) is the prior probability of class.
- P(x|c) is the probability of predictor given class.
- P(x) is the prior probability of predictor, where C and X are two events.

In naïve Bayes classifier, probability theory is used to find the unclassified instance.

10.3.3.2 Random Forest

Random forest is an ML algorithm that falls under supervised learning. It is utilized for both regression and classification. It makes more decision trees from the available data and gets the prediction of each decision tree [15]. The final selection of the solution or the final prediction is done by voting. The following steps clearly explain the random forest algorithm:

- Select the random sample from the given dataset.
- A decision tree is constructed for every sample, and the prediction of each decision tree is noted.
- Voting is carried out from the prediction of each decision tree.
- The most voted prediction result is considered as final.

10.3.3.3 Decision Tree

A decision tree is an ML algorithm that is used for classification and regression [17]. It falls under nonparametric supervised learning. The main objective of the decision tree is to make a model that predicts the value of variables. The prediction is based on simple decision rules. It is popular due to its adaptability to produce logical rules.

10.3.3.4 AdaBoost Algorithm

AdaBoost is an ensemble method that iteratively constructs a classifier. In every iteration, it calls a simple learning algorithm (the weak learner) that returns a classification. The final classification is decided by a weighted vote of the weak classifiers, where each weight is proportional to the correctness of the corresponding weak classifier. This incremental process of combining weak classifiers weighed by their performance is called *boosting*. Weak classifiers need only to be slightly better than a random guess, which lends great flexibility to the design of the weak classifier (or feature) set. If there is no particular a priori knowledge available on the domain of the learning problem, small decision trees or decision stumps (decision trees with two leaves) are often used. A decision stump is defined by three parameters: the index j of the attribute that it cuts, the threshold μ of the cut, and the sign of the decision.

In this algorithm, weak classifiers are selected iteratively by some candidate weak classifiers and combined linearly to form a strong classifier for classifying the network data. Let $H = \{h_f\}$ be the set of constructed weak classifiers. Let the set of training sample data be $(x_1, y_1), \dots (xi, yi), \dots (xn, yn)$, where xi denotes the ith feature vector; $yi \in +1, -1$ be the label of the ith feature vector, denoting whether the feature vector represents a normal behavior or not; and n be the size of the dataset.

This algorithm runs for T rounds. Each sample i is assigned a weight $wi(t)$ at any round t. Initially, all weights are set equally, but during the execution of the algorithm, these weights are redistributed to manipulate the selection process. In every round t, the performance of each single weak classifier hj on all n training samples is assessed. The performance is measured by the weighted error defined as:

$$\varepsilon_j = \sum_{n}^{i=1} w_i(t) I[y_i \neq h_j(x_i)] \tag{10.2}$$

$$\text{where } I_\gamma = \begin{cases} 1 & \gamma = True \\ 0 & \gamma = False \end{cases}.$$

At the end of each round, the classifier h with the lowest error rate εt based on Equation (10.2) is selected and stored as the best classifier ht of round t. Then, a confidence αt is computed as:

$$\alpha_t = \frac{1}{2} \log\left(\frac{1-\varepsilon_t}{\varepsilon_t}\right) \tag{10.3}$$

and can be interpreted as the quality of hypothesis ht, the lower error rate εt, and the higher confidence αt. Intuitively, αt measures the importance that is assigned to ht. Note that $\alpha t > 0$ if $\varepsilon t > \frac{1}{2}$ and that αt gets larger as εt gets smaller. Finally, the weights are redistributed and normalized as follows:

$$w_i(t+1) = \frac{w_i(t) \exp\left(-\alpha_t y_i h_t\ (x_i\ )\right)}{z_t} \tag{10.4}$$

By increasing the weights of samples that were misclassified by ht favors, in the next round, these difficult samples are handled correctly. The effect of this rule is to increase the weight of examples misclassified by ht and to decrease the weight of correctly classified examples. Thus, the weight tends to concentrate on "hard" examples. Zt denotes the normalization factor which ensures the sum of all weights to be +1. The normalization factor is defined as:

$$Z_t = \sum_{k=1}^{n} \exp(-\alpha_t y_i h_t(x_k)) \tag{10.5}$$

The final hypothesis H is a weighted majority vote of the T selected weak classifiers where αt is the weight assigned to hypothesis ht:

$$H(x) = sign\left(\sum_{t=1}^{T} \alpha_t h_t(x)\right) \tag{10.6}$$

The pseudocode is:
Require (x_1, y_1), ... (xn, yn); $xi \in X$, $yi \in \{+1, -1\}$
Initialize weights $w_1\ (i) = 1/n$
For all t such that $0 \leq t \leq T$ do
Find the minimum classification error $ht = \min_{hj \in H} \varepsilon j$

if $\varepsilon j \geq {}^{1}$, then
Reinitialize the weights to $1/n$
Go to step 3
end if
Compute the confidence αt
Update the weights $w(t+1)$
Normalize weight of each sample
end for
Output the strong classifier $H(x)$.

10.3.4 Performance Analysis

The following parameters are used to evaluate the efficiency of the algorithm:

$$\text{Accuracy} = \frac{\text{TP} + \text{TN}}{\text{TP} + \text{TN} + \text{FP} + \text{FN}}$$

$$\text{Sensitivity} = \frac{\text{TP}}{\text{FN} + \text{TP}}$$

$$\text{Specificity} = \frac{\text{TN}}{\text{FP} + \text{TN}}$$

$$\text{Precision} = \frac{\text{TP}}{\text{FP} + \text{TP}}$$

$$\text{F-measure} = \frac{2\text{TP}}{2\text{TP} + \text{FP} + \text{FN}}$$

The parameters TP (true positive), TN (true negative), FP (false positive), and FN (false negative) are taken from the confusion matrix generated during the classification.

10.3.5 Result

The efficiency-measuring parameters for various algorithms are studied. It is observed that there is a slight enhancement in accuracy and diminution of error rate while using the AdaBoost algorithm. RF algorithm has the maximum precision rate. The LR and linear model have the highest F-measure and specificity values when compared to the other ML algorithms. The results are tabulated in Table 10.2. Figures 10.2–10.7 give a graphical representation of the results obtained.

10.4 Conclusion

Cardiac disease is one of the major diseases worldwide. Using ML techniques, it is possible to predict heart disease at an early stage, through which the mortality rate due to heart disease can be reduced. The prediction has to be accurate and must have less error rate, and using a suitable algorithm is hence an important factor. The AdaBoost algorithm increases

Table 10.2 Comparison of parameters for different algorithms.

Model	Accuracy	Precision	F-measure	Sensitivity	Specificity	Error
Naïve Bayes	75.8	90.5	84.5	79.8	60	24.2
Logistic regression	82.9	88.8	91.6	94.9	20	14.9
Decision tree	70	76.92	68.9	62.5	80	30
Random forest	87.09	100	86.6	76.47	100	12.91
Linear model	85.1	88.8	91.6	94.9	20	14.9
AdaBoost (proposed)	90.32	92.30	88.88	85.71	94.11	9.68

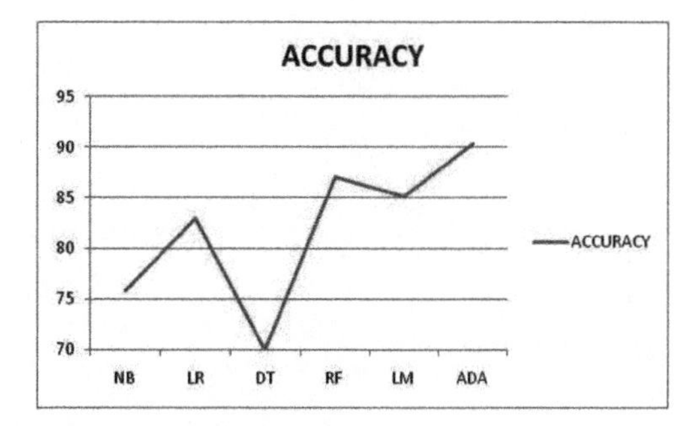

Figure 10.2 Accuracy graph of various algorithms.

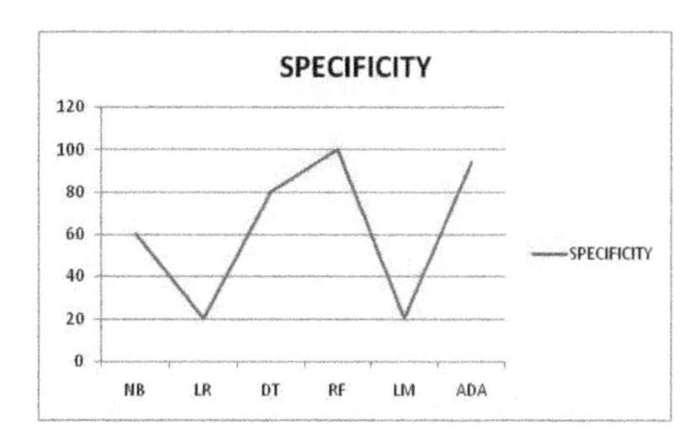

Figure 10.3 Specificity graph of various algorithms.

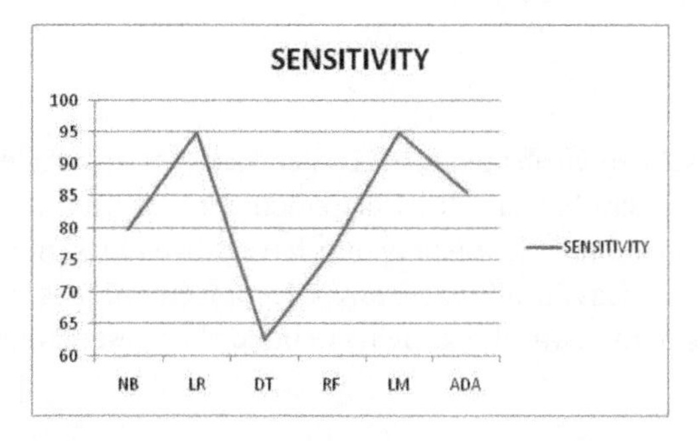

Figure 10.4 Sensitivity graph of various algorithms.

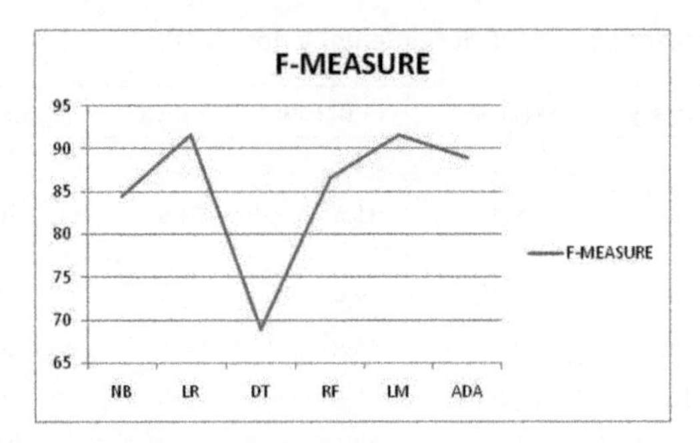

Figure 10.5 F-measure graph of various algorithms.

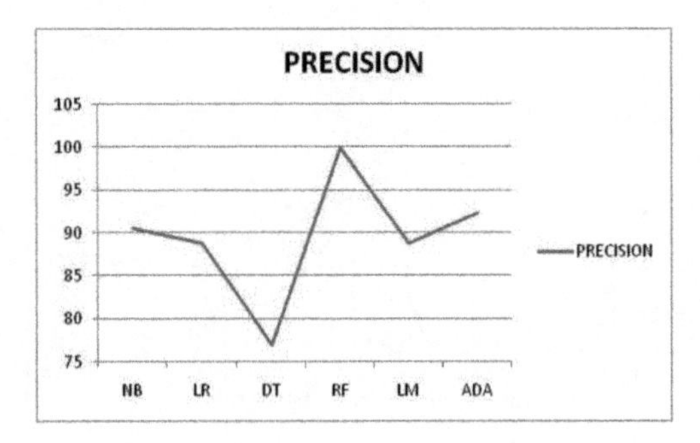

Figure 10.6 Precision graph of various algorithms.

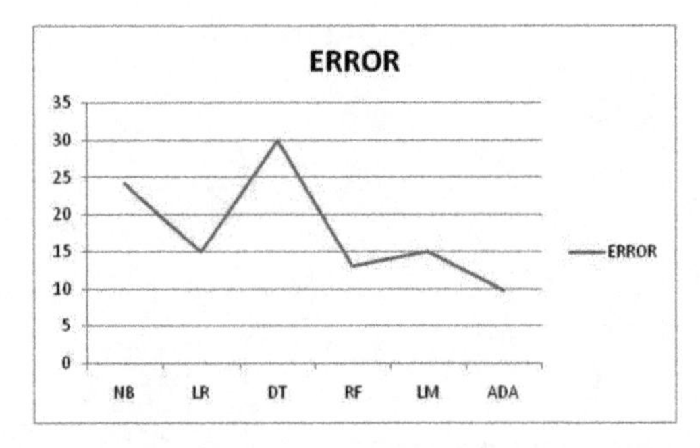

Figure 10.7 Error graph of various algorithms.

the weight of the misclassified elements, which decreases the error rate and increases accuracy. Further improvement in heart disease prediction is required to improve accuracy by using different ML algorithms. Neural networks have proven to be more efficient in prediction and can be used for classification. Feature selection techniques can be created to get a more extensive view of the critical highlights to build the presentation of coronary illness forecast.

References

[1] F. Sabahi, "Bimodal fuzzy analytic hierarchy process (BFAHP) for coronary heart disease risk assessment," *J. Biomed. Informat.*, vol. 83, pp. 204–216, July 2018. doi:10.1016/j.jbi.2018.03.016.

[2] K. Uyar and A. Ilhan, "Diagnosis of heart disease using genetic algorithm based trained recurrent fuzzy neural networks," *Procedia Comput. Sci.*, vol. 120, pp. 588–593, 2017.

[3] P.S. Kumar, D. Anand, V.U. Kumar, D. Bhattacharyya, and T.-H. Kim, "A computational intelligence method for effective diagnosis of heart disease using genetic algorithm," *Int. J. Bio-Sci. Bio-Technol.*, vol. 8, no. 2, pp. 363–372, 2016.

[4] R. Wagh and S.S. Paygude, "CDSS for heart disease prediction using risk factors," *Int. J. Innov. Res. Comput.*, vol. 4, no. 6, pp. 12082–12089, June 2016.

[5] A. Gavhane, G. Kokkula, I. Pandya, and K. Devadkar, "Prediction of heart disease using machine learning," In *Proceedings of the 2nd International Conference on Electronics, Communication and Aerospace Technology (ICECA)*, March 2018; pp. 1275–1278.

[6] V. Krishnaiah, G. Narsimha, and N. Subhash, "Heart disease prediction system using data mining techniques and intelligent fuzzy approach: A review," *Int. J. Comput. Appl.*, vol. 136, no. 2, pp. 43–51, 2016.

[7] J. Nahar, T. Imam, K.S. Tickle, and Y.-P.-P. Chen, "Computational intelligence for heart disease diagnosis: A medical knowledge driven approach," *Expert Syst. Appl.*, vol. 40, no. 1, pp. 96–104, 2013. doi:10.1016/j.eswa.2012.07.032.

[8] T. Mahboob, R. Irfan, and B. Ghaffar, "Evaluating ensemble prediction of coronary heart disease using receiver operating characteristics," In *Proceedings of the Internet Technologies and Applications (ITA)*, September 2017; pp. 110–115.

[9] R. Das, I. Turkoglu, and A. Sengur, "Effective diagnosis of heart disease through neural networks ensembles," *Expert Syst. Appl.*, vol. 36, no. 4, pp. 7675–7680, May 2009. doi:10.1016/j.eswa.2008.09.013.

[10] Y.E. Shao, C.-D. Hou, and -C.-C. Chiu, "Hybrid intelligent modeling schemes for heart disease classification," *Appl. Soft Comput. J.*, vol. 14, pp. 47–52, January 2014. doi:10.1016/j.asoc.2013.09.020.

[11] M.S. Amin, Y.K. Chiam, and K.D. Varathan, "Identification of significant features and data mining techniques in predicting heart disease," *Telematics Inform.*, vol. 36, pp. 82–93, March 2019 .[Online]. Available: https://linkinghub.elsevier.com/retrieve/pii/S0736585318308876.

[12] T. Vivekanandan and N.C.S.N. Iyengar, "Optimal feature selection using a modified differential evolution algorithm and its effectiveness for prediction of heart disease," *Comput. Biol. Med.*, vol. 90, pp. 125–136, November 2017.

[13] O.W. Samuel, G.M. Asogbon, A.K. Sangaiah, P. Fang, and G. Li, "An integrated decision support system based on ANN and Fuzzy_AHP for heart failure risk prediction," *Expert Syst. Appl.*, vol. 68, pp. 163–172, February 2017.

[14] L. Baccour, "Amended fused TOPSIS-VIKOR for classification (ATOVIC) applied to some UCI data sets," *Expert Syst. Appl.*, vol. 99, pp. 115–125, June 2018. doi:10.1016/j.eswa.2018.01.025.

[15] S. Mohan, C. Thirumalai, and G. Srivastava, "Effective heart disease prediction using hybrid machine learning techniques," *IEEE Access*, vol. 7, pp. 81542–81554, 2019.

[16] M. Gandhi and S.N. Singh, "Predictions in heart disease using techniques of data mining," In *Proceedings of the International Conference on Futuristic Trends on Computational Analysis and Knowledge Management (ABLAZE)*, February 2015; pp. 520–525.

[17] J. Thomas and R.T. Princy, "Human heart disease prediction system using data mining techniques," In *Proceedings of the International Conference on Circuit, Power and Computing Technologies (ICCPCT)*, March 2016; pp. 1–5.

11

Geographic Information Systems and Confidence Interval with Deep Learning Techniques for Traffic Management Systems in Smart Cities

Prisilla Jayanthi

The Airports Authority of India Ltd.

11.1 Introduction

A traffic system involves the complex task of integrating information and communication technology (ICT) and the Internet of things (IoT). A traffic control algorithm works on scheduling algorithms and monitors traffic movement. The communication is carried out using single-cast, broadcast, multicast, and other cast communication methods [1]. This involves centralized and decentralized architectures. Traffic management plays a vital role in structuring SCs. In this chapter, a few case studies are shown of road accidents caused by various reasons such as environmental factors, drunken driving, poor education, and age. SC models and traffic congestion prediction algorithms are discussed.

11.2 Traffic Management for Smart Cities

The system of smart traffic management is where centrally controlled traffic signals and sensors regulate the flow of traffic through a city in response to demand. Traffic management systems (TMSs) for SCs encompass wide components that include the collection and processing of real-time traffic data, traffic monitoring platformsand traffic control systems for traffic signs, and digital road signage systems. The primary purpose of SC is to address rampant urban congestions. A real-time dynamic TMS intelligently navigates automobiles to optimize traffic flow on the roads and to solve gridlocks in cities.

11.2.1 Smart City Model

SC uses ICT to enhance livability, sustainability, and workability. It integrates with IoTs to manage city assets.

The smart town/city model was proposed based on Giffinger's approach titled after Rudolf Giffinger. The six characteristics in Figure 11.1 define many factors. The six different characteristics of the smart town are shown in Table 11.1 with assigned factors and the corresponding

Sensor Data Analysis and Management: The Role of Deep Learning, First Edition. Edited by A. Suresh, R. Udendhran, and M.S. Irfan Ahmed.
© 2021 John Wiley & Sons, Ltd. Published 2021 by John Wiley & Sons, Ltd.

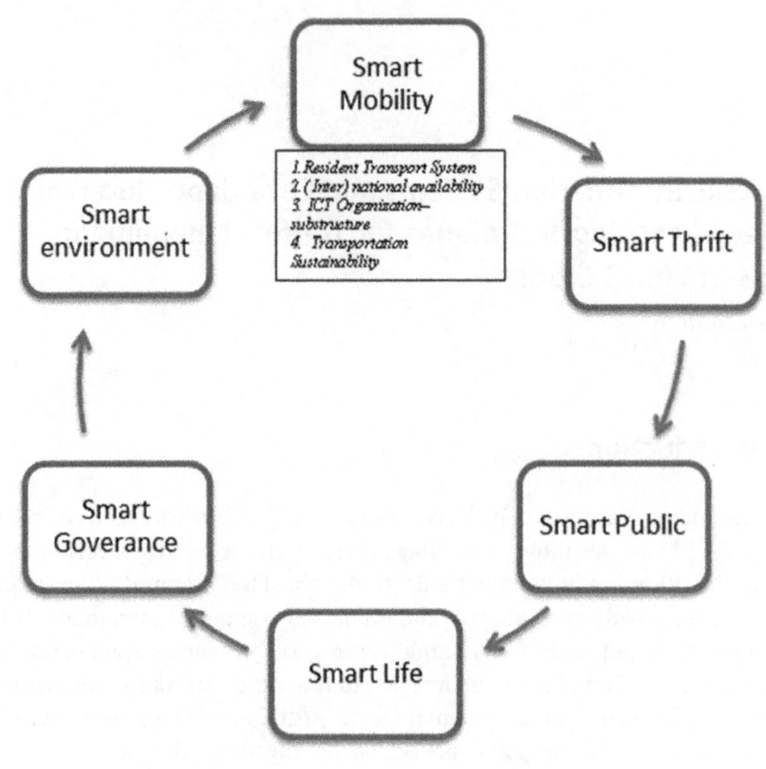

Figure 11.1 The smart town model.

Table 11.1 Smart town's characteristics, factors, and measures [2].

Characteristics	Medium-sized cities (100 000–500 000 inhabitants)		Larger cities (300 000–1 million inhabitants)	
	Factors	**Measure**	**Factors**	**Measure**
Smart thrift (competitiveness)		15		15
	Inventive spirit	3	Inventive spirit	3
	Entrepreneur	3	Entrepreneur	3
	Remunerative trademarks	1	City image	2
	Production	3	Production	3
	Flexibility of trader market	3	Trader market	2
	Global embeddedness	2	Global integration	2
Smart mobility (transport and ICT)		11		13
	Resident accessibility	3	Resident transport system	2
	(Inter)national availability	1	(Inter)national availability	1

(*Continued*)

Table 11.1 (*Continued*)

Characteristics	Medium-sized cities (100 000–500 000 inhabitants)		Larger cities (300 000–1 million inhabitants)	
	Availability of IT substructure	3	ICT organization-substructure	4
	Sustainability of the transport system	4	Transportation sustainability	6
Smart environment (natural resources)		10		10
	Environment circumstances	2	Reduced or no air pollution	4
	No air pollution	3	Ecological awareness	4
	Environmental awareness	3	Sustaining source supply management	2
	Sustainable resource management	2		
Smart public (social and human capital)		11		11
	Qualification level	2	Edification	1
	Lifetime learning	3	Lifetime learning	2
	Indigenous plurality	2	Ethnic plurality	3
	Broad-mindedness	4	Broad-mindedness	5
Smart life (quality of life)		25		31
	Traditional facilities	3	Cultural and leisure facilities	6
	Well-being conditions	6	Well-being conditions	5
	Individual security	2	Individual security	3
	Housing quality	3	Accommodation standard	4
	Edification facilities	5	Schooling facilities	4
	Tourist attractions	1	Tourist attractions	5
	Economic welfare	5	Social cohesion	4

number of measures. The factors and measures of a smart town are determined by its total population. Big cities have many more measures to describe smart life and smart mobility.

SC needs smart management and administration, and few measures need to be implemented to refer it as an "ultramodern urban area." These measures include implementing a real-time system to display the timing for public transport at each station, green transport systems (public pay-and-use bike-sharing systems), intelligent parking systems, intelligent traffic lights systems, public lights, and intelligent sensors. The concept of bike sharing provides an initiative for mobility that helps in CO_2 emissions reduction. This bike sharing concept has been already implemented in countries such as the USA, UK, etc. The concept has proven to have a positive impact on citizens' daily lives, providing a mode of travel that combines affordable transport on demand and improvements in traffic flow. In India, car sharing (cab) has been implemented, which helps tourists travel through the city in a faster and safe manner [2].

Boyd Cohen proposed "smart cities wheel" as another smart town model. The model (Figure 11.2) has all six features that were defined by Giffinger. It was modeled to support smart town strategies, develop baselines, and track the progress transparently [3].

11.2.2 Smart City–Traffic Management

Traffic management is a complex task with large risks and the responsibility of controlling the traveling citizens on a day-to-day basis. It involves governing and controlling static and

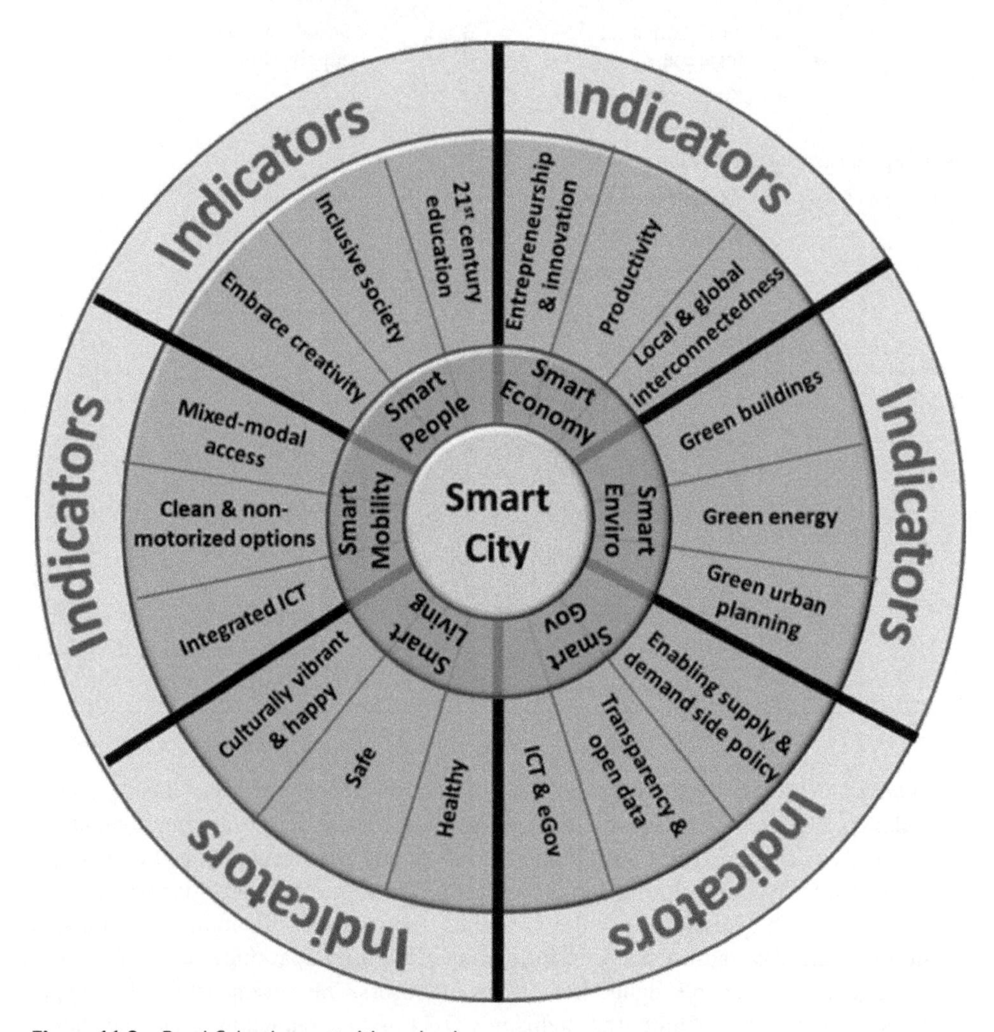

Figure 11.2 Boyd Cohen's smart cities wheel.

moving traffic, including pedestrians and two-wheeler or four-wheeler vehicles. Its target is to provide for the safe, orderly, and efficient movement of persons and goods, and to protect and enhance the quality of the local environment on and adjacent to traffic facilities. The traffic study includes an inventory of roads and statistical methods along with traffic volume and composition, origin and destination, speed, travel time and delay, accidents, and parking. Several aspects of traffic management include signs and delineation, pedestrian paths, intersection points, traffic signals, road capacity, parking, and roadway public lights. All these tasks need to be interconnected with nodes and communicated on time, which necessitates smarter cities integrating IoT and ICT.

Devi and Neetha [4] proposed a congestion prediction algorithm (Figure 11.3) based on machine learning that used logistic regression, providing accurate and early prediction of the traffic congestion for a given static road network. The inputs for this algorithm are five different attributes—vehicle number (VID), time stamp (TS) at which the data are collected, vehicle speed (VS), X and Y coordinates of the vehicle at a particular time (X and Y at t1), and congestion (C) in the path in which it is traveling. The last attribute is a label (threshold) that is assigned as "congested" using the algorithm. In another case study, Khanna et al. [5] proposed a novel intelligent TMS (Figure 11.4) for SCs that facilitate wireless sensor networks, IoTs, cloud computing, and data analytics. It was successful in predicting traffic congestion levels, and the system reports accidents and the impact of flow of traffic of that region at that hour.

The taxonomy table in Table 11.2 describes all the entities used in the derived mathematical model $\omega = ((D/\text{speed} - T) \times 100)/(D/\text{speed})$ suggested by Khanna et al. [5]. When road capacity (RC) is greater than the threshold value, the entry of all heavy vehicles into that zone is prevented by the road traffic management controller. Any vehicle in the affected zone is directed to stop until they receive further directions from the central controller. At $X = 3$, the traffic light goes green 10 seconds before the arrival of any emergency vehicle at the intersection.

11.3 Elements Responsible for Road Accidents–Causing Traffic

Road accidents often lead to loss of human life. Human errors are often the reason for the occurrence of accidents and crashes. A few are listed here: overspeed driving, drinking and driving, distractions to driver, jumping red light signal, etc. A few case studies are discussed in this chapter.

In a case study, Rolison et al. [6] discuss the factors causing road accidents, which include insufficient experience in driving, inept maneuvering skills, and risk-taking behaviors, all of which are commonly associated with young drivers. On the other hand, visual, cognitive, and mobility impairments have been associated with collisions in older drivers. Figure 11.5 lists the percentage of police officers and the public who determine each cause as per the age and gender of the driver in hypothetical road accident scenarios. The color coding identifies the most frequent (red) to the least frequent (green) factors separately for each driver's age and gender.

Table 11.2 Taxonomy table [5].

Symbol	Denotation
Γ	Road sensors
V	Automobile node
r	Road
R	Path
A	Fuel consumption
β	Traffic flow density
Φ	Moving road traffic velocity
T	Average waiting time
H	Total waiting time
Ψ	Automobile type
Δ	Automobile state
N	Number of intersections
E	Automobile priority
Θ	Optimum vehicle speed
RC	Road capacity
RS	Route selection function
F	Travel cost function
W	Road traffic management controller
X	Traffic signal
Ω	Rush-hour traffic flow percentage

Daytime collisions happen for various reasons. Police officers and the public observed drugs or alcohol as a factor contributing to collisions at night (%police = 76; %public = 58) as compared to the evening (%police = 62; %public = 41) and during daytime (%police = 44; %public = 31). Police officers (8% vs. 1%) stated that poor visibility during the evening (as compared to daytime) was one of the reasons for collisions—as compared to the public's opinion (25% vs. 5%). In comparison to the evening, night-time accidents were more frequently determined by police officers (16%) as compared to the public (33%). Inexperience; consumption of pills, drugs, or alcohol; and exceeding the speed limit were strongly associated with driver's age in participants' views regarding road accident records, as shown in Table 11.3. Failure to see properly, loss of control, distraction, and failure to judge another person's path or speed were some frequently reported issues in the road accident records.

Sami et al. [7] include education, age, and gender in their study of factors that are responsible for road accidents. They divided the education level into three sets: low, medium, and high levels. The results in Table 11.4 show that low-level education causes higher mortality rate in accidents among males, but this is not the same with females. In all three sets, the highest rate of mortality is associated with the age range of 30–49 years among males. The

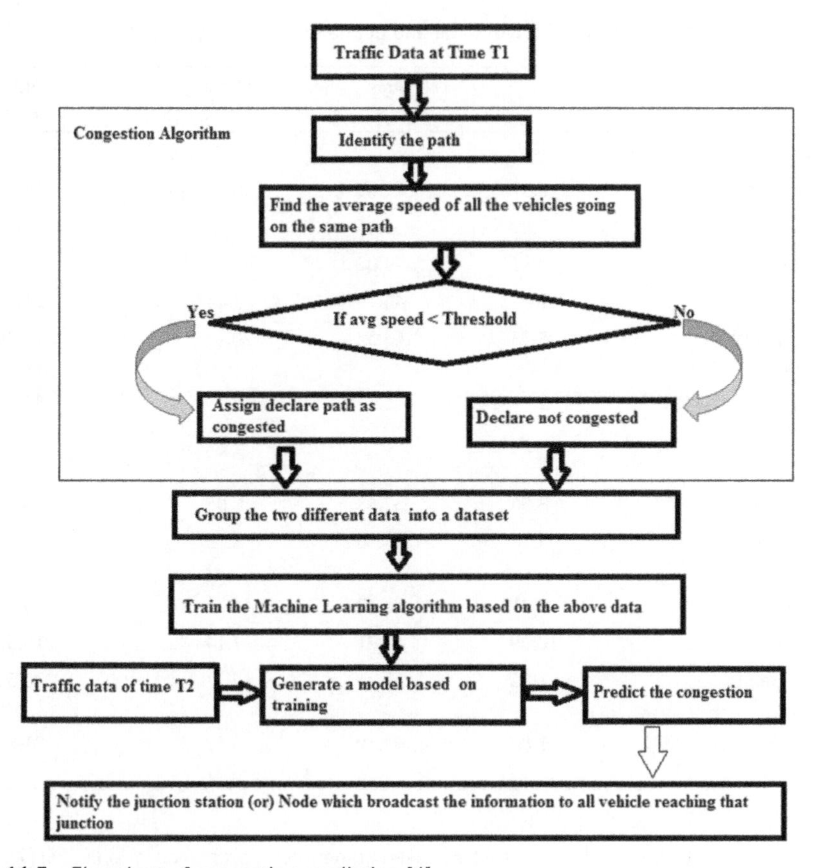

Figure 11.3 Flowchart of congestion prediction [4].

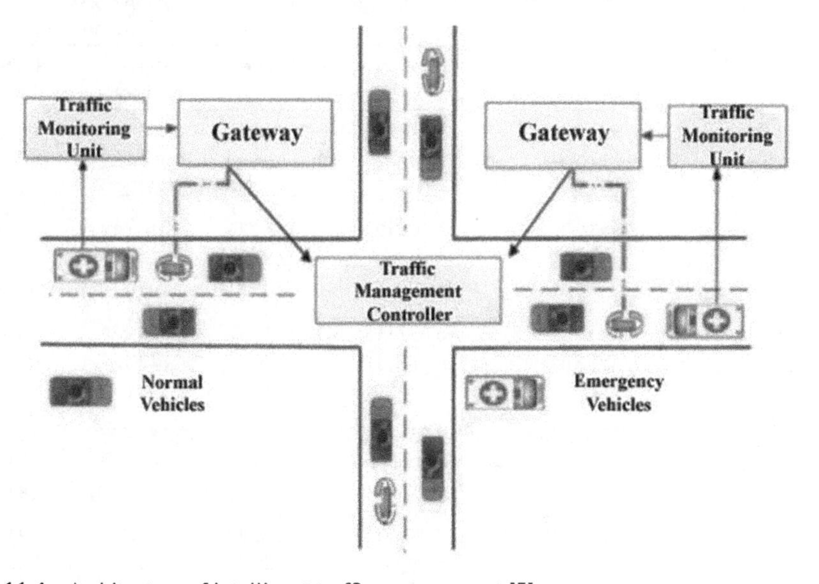

Figure 11.4 Architecture of intelligent traffic management [5].

Table 11.3 Statistical analysis of driver's age and gender in road accident occurrence [6].

	Sex	Age	Middle vs. young age	Older vs. middle age
Police				
Pills or alcohol	0.67	21.31	0.89	16.33
Extreme speed	0.82	65.46	2	29.12
Inexperience	0.82	79.02	44	0
Disturbance	7	46.19	5.44	21.16
Physical health condition	5.33	63.52	20.17	19.30
Eyesight (uncorrected or defective)	2	75.17	8	35.10
Public				
Pills or alcohol	3.56	37.68	8.05	14.29
Extreme unnecessary speed	3.77	58.36	8.05	8.13
Inexperience	9.31	88.48	46	0.67
Disturbance	10.29	45.21	7.14	19.59
Physical health condition	1.92	59.21	17	20.57
Eyesight (uncorrected or defective)	1.80	67.19	3.57	31.84

Table 11.4 Road traffic accident mortalities based on gender, age, and education (*n*, %) [7].

Age	Population and mortality	Uneducated	Low-level education		Proper-level education	
			Elementary	Junior	High school	University
10–19 M	Pop.	9218 (0.24)	74 647 (1.93)	179 988 (4.65)	205 290 (5.31)	16 248 (0.42)
	Traf. Mort.	17 (0.93)	30 (1.64)	55(3.00)	50 (2.73)	5 (0.27)
F	Pop	8498 (0.22)	78 708 (2.03)	148 229 (3.83)	210 618 (5.45)	18 466 (0.48)
	Traf. Mort.	3 (0.16)	4 (0.22)	4 (0.22)	10 (0.55)	0 (0)
20–29 M	Pop.	14 874 (0.38)	56 837 (1.47)	147 081 (3.80)	198 439 (5.13)	98 925 (2.56)
	Traf. Mort.	37 (2.02)	46 (2.51)	153 (8.36)	136 (7.43)	54 (2.95)
F	Pop.	16 445 (0.43)	89 831 (2.32)	98 138 (2.54)	184 677 (4.77)	122 704 (3.17)
	Traf. Mort.	4 (0.22)	7 (0.38)	16 (0.87)	29 (1.58)	18 (0.98)
30–39 M	Pop.	17 831 (0.46)	67 990 (1.76)	92 509 (2.39)	83 359 (2.16)	49 940 (1.29)
	Traf. Mort.	17 (0.93)	42 (2.29)	85 (4.64)	51 (2.79)	22 (1.20)

(*Continued*)

Table 11.4 (*Continued*)

Age	Population and mortality	Uneducated	Low-level education		Proper-level education	
			Elementary	Junior	High school	University
F	Pop. Traf. Mort.	37 269 (0.96) 5 (0.27)	107 549 (2.78) 8 (0.44)	64 878 (1.68) 7 (0.38)	58 751 (1.52) 7 (0.38)	34 512 (0.89) 11 (0.60)
40–49 M	Pop. Traf. Mort.	29 471 (0.76) 30 (1.64)	63 284 (1.64) 58 (3.17)	49 154 (1.27) 56 (3.06)	57 700 (1.49) 33 (1.80)	32 490 (0.84) 6 (0.33)
F	Pop. Traf. Mort.	64 065 (1.66) 13 (0.71)	74 818 (1.93) 18 (0.98)	28 431 (0.74) 11 (0.60)	40 385 (1.04) 8 (0.44)	12 954 (0.33) 2 (0.11)
50–59 M	Pop. Traf. Mort.	49 569 (1.28) 30 (3.39)	62 519 (1.62) 72 (3.93)	17 681 (0.46) 21 (1.15)	28 946 (0.75) 31 (1.69)	19 012 (0.49) 9 (0.49)
F	Pop. Traf. Mort.	103 152 (2.67) 36 (1.92)	48 239 (1.25) 15 (0.82)	10 463 (0.27) 5 (0.27)	16 058 (0.42) 9 (0.49)	6492 (0.17) 1 (0.05)

M = male, F = female, Pop. = population, Traf. = traffic, Mort. = Mortality

result shows that illiterates had fatalities with age above 64 years, and less (un)educated groups with ages within 20–29 years. Figure 11.6 shows the accident mortality of each educational level per 100 000 people. The mortality per 100 000 people was to be found inversely proportional to the education level. However, mortality was marginally higher with respect to elementary and junior high schools. Mortality rate in the age group 20–29 years was higher than in other age groups.

Jalilian et al. [8] considered that environmental factors play a significant role in road accident occurrence. COM114 is the system that collects data from road traffic accidents (RTAs), and where the database of all traffic crashes gets registered. The experts collect information about traffic crashes, including the persons, vehicles, and environmental conditions involved (Table 11.5). The data in the COM114 form are collected in two sections—general and specific information. In their road accident study, the time, location, road status, weather conditions, lighting conditions, vehicle characteristics, behavior of the road user (driver, passenger, and pedestrian), and the causes of the RTAs are recorded.

The adjusted chance of injurious or fatal RTAs in avenues was 9.74. In sidetracks, it was 3.54 times more than in highways. The adjusted chance of injurious or fatal traffic accidents occurring in the evening was 2.31 times more than in daytime; and, on a cloudy day, it was 2.60 times more than on a clear day (Table 11.6). Table 11.6 describes the estimation of the occurrence of RTAs that lead to death or injury caused due to environmental factors. The highest frequency of accidents had occurred on clear days, which might have been caused by not driving with caution.

Police	Male drivers			Female drivers		
	Younger	Middle age	Older	Younger	Middle age	Older
Other driver (third party)	3.90	6.49	3.90	2.60	3.90	2.60
Unfamiliar with road (layout, route)	1.30	2.60	2.60	3.90	0.00	6.49
Drugs or alcohol	80.52	74.03	45.45	63.64	70.13	29.87
General driving ability (skills)	6.49	2.60	9.09	7.79	9.09	10.39
Excessive speed	80.52	72.73	32.47	72.73	59.74	24.68
Inexperience	61.04	3.90	2.60	62.34	6.49	7.79
Dangerous driving (peer pressure, showing off)	31.17	9.09	5.19	11.69	1.30	2.60
Distraction (phone, friends, kids, outside)	68.83	55.84	19.48	88.31	67.53	22.08
Driver error (poor judgement)	14.29	11.69	36.36	15.58	22.08	40.26
Road conditions (road layout, road hazard)	20.78	28.57	20.78	28.57	29.87	28.57
Inattention (concentration)	15.58	20.78	25.97	20.78	27.27	15.58
Careless, reckless or in a hurry	12.99	11.69	9.09	15.58	11.69	9.09
Vehicle defects (mechanical failure)	22.08	16.88	12.99	14.29	14.29	9.09
Weather	25.97	29.87	23.38	20.78	31.17	28.57
Overconfidence	7.79	3.90	1.30	1.30	3.90	0.00
Failed to look properly (poor observations)	6.49	9.09	10.39	5.19	5.19	11.69
Traffic	2.60	2.60	2.60	3.90	1.30	3.90
Fatigue	12.99	32.47	19.48	15.58	28.57	19.48
Driving too slow for conditions or slow vehicle	0.00	0.00	0.00	0.00	0.00	3.90
Slow driver reaction	0.00	0.00	20.78	0.00	2.60	22.08
Medical condition (physical impairment, medication)	10.39	35.06	77.92	10.39	23.38	67.53
Poor visibility	5.19	5.19	9.09	6.49	7.79	16.88
Eyesight (uncorrected or defective)	0.00	7.79	57.14	0.00	6.49	53.25
Dazzling light (headlights or sunlight)	1.30	5.19	2.60	0.00	5.19	2.60
Nervous or uncertain (hesitation, confusion, lack of confidence)	0.00	1.30	11.69	2.60	9.09	15.58

Figure 11.5 Percentage of road accidents segregated in terms of cause, gender, and age [6].

Public

	Male drivers			Female drivers		
	Younger	Middle age	Older	Younger	Middle age	Older
Other driver (third party)	9.80	9.80	11.76	4.90	10.78	10.78
Unfamiliar with road (layout, route)	1.96	0.00	2.94	0.98	1.96	1.96
Drugs or alcohol	74.51	50.00	22.55	54.90	38.24	19.61
General driving ability (skills)	1.96	0.98	1.96	2.94	0.98	2.94
Excessive speed	69.61	54.90	23.53	52.94	46.08	23.53
Inexperience	38.24	0.00	0.98	49.02	5.88	2.94
Dangerous driving (peer pressure, showing off)	28.43	4.90	0.00	5.88	3.92	0.98
Distraction (phone, friends, kids, outside)	50.00	34.31	11.76	71.57	53.92	15.69
Driver error (poor judgement)	8.82	15.69	14.71	8.82	9.80	14.71
Road conditions (road layout, road hazard)	16.67	30.39	22.55	20.59	23.53	20.59
Inattention (concentration)	19.61	25.49	24.51	18.63	36.27	24.51
Careless, reckless or in a hurry	16.67	15.69	9.80	12.75	16.67	9.80
Vehicle defects (mechanical failure)	11.76	16.67	14.71	13.73	16.67	15.69
Weather	28.43	33.33	34.31	27.45	36.27	24.51
Overconfidence	3.92	2.94	1.96	2.94	0.00	0.00
Failed to look properly (poor observations)	0.98	0.98	0.98	0.00	0.98	1.96
Traffic	2.94	3.92	4.90	1.96	4.90	4.90
Fatigue	18.63	32.35	30.39	17.65	32.35	31.37
Driving too slow for conditions or slow vehicle	0.00	0.00	2.94	0.00	0.00	0.98
Slow driver reaction	0.00	0.00	19.61	0.00	0.98	18.63
Medical condition (physical impairment, medication)	2.94	19.61	43.14	1.96	14.71	41.18
Poor visibility	17.65	21.57	21.57	17.65	21.57	27.45
Eyesight (uncorrected or defective)	0.98	3.92	37.25	0.98	4.90	30.39
Dazzling light (headlights or sunlight)	2.94	1.96	3.92	1.96	2.94	5.88
Nervous or uncertain (hesitation, confusion, lack of confidence)	0.98	2.94	7.84	4.90	4.90	6.86

Figure 11.3 (Continued)

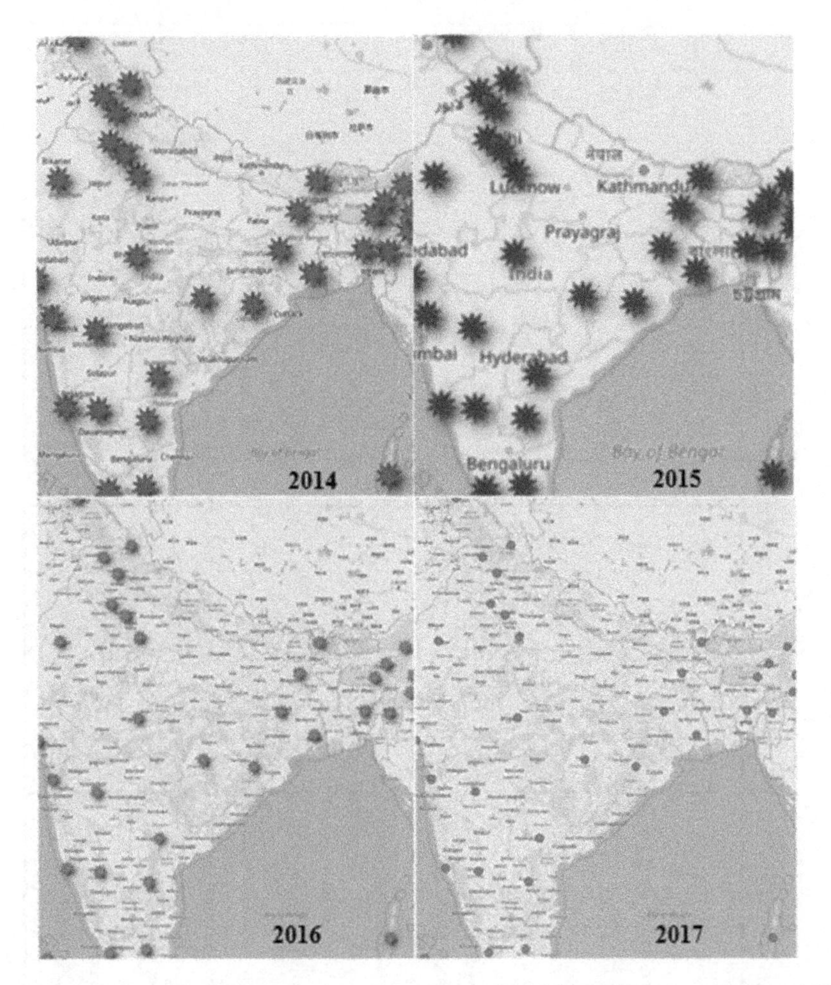

Figure 11.6 QGIS results for state-wise road accidents for (a) 2014, (b) 2015, (c) 2016, and (d) 2017.

11.4 Analysis of Road Accidents Using Geographic Information Systems

A framework used for capturing, gathering, storing, managing, manipulating, and analyzing all types of geographical data is referred to as a geographic information system (GIS). GIS integrates many types of data. It analyzes spatial location and organizes layers of information into visualizations using maps and 3D scenes. GIS provides deeper insights into data, such as patterns, relationships, and situations—helping users make smarter decisions. In this study, GIS has been implemented for analyzing the road accidents in different states of India for the years 2014–2017. It helps to understand the exact locations where the accidents occur, and a lot more information is gained such as the feature density

Table 11.5 Environmental factors responsible for road accidents [8].

Variable	Accidents causing		Total	P-value
	Injury or death	No injuries		
Light condition				
Day	970 (58.4)	689 (41.5)	1656 (100)	
Night	350 (65.3)	186 (34.7)	536 (100)	0.003
Daylight	11 (57.8)	8 (42.1)	19 (100)	
Sunset	59 (73.7)	21 (26.2)	80 (100)	
Total	1390 (60.5)	904 (39.4)	12294 (100)	
Driver's visual condition				
Visual obstruction (yes)	22 (40.7)	32 (59.2)	54 (100)	
Visual obstruction (no)	1260 (59.0)	873 (40.9)	2133 (100)	0.007
Total	1282 (58.6)	905 (41.3)	2187 (100)	
Accident's position				
Road	1050 (59.5)	712 (40.4)	1762 (100)	
Verge	59 (48.3)	63 (51.6)	122 (100)	<0.05
Edge of the road	170 (57.2)	127 (42.7)	297 (100)	
Total	1279 (58.6)	902 (41.3)	2181 (100)	
Sort of the track				
Highway	3 (21.4)	11 (78.6)	14 (100)	
Path	688 (73.2)	251 (26.8)	939 (100)	
Main road	471 (50.5)	462 (49.5)	933 (100)	0.001
Sidetrack	333 (54.5)	194 (45.5)	427 (100)	
Total	1395 (60.3)	918 (39.7)	2313 (100)	

of a given space and what is happening inside an area of interest. The data are collected from an open-source portal (www.data.gov.in).

Tables 11.7a and 11.7b represent the data for different states in India for the years 2014–2017 where the road accidents have occurred, calculated per 100,000 population. Table 11.8 shows the different types of accidents such as fatal, minor, non-injury, or grievous injury. Figure 11.6 represents the graph for Tables 11.7a and 11.7b for separate years obtained by QGIS 3.10.0. Figure 11.7 shows the results for Table 11.7 obtained by QGIS 3.10.0 for fatal and grievous injury accidents and minor injury and non-injury accidents, respectively. Table 11.7b provides information on the total number of persons injured in road accidents state-wise per 100,000 population for the years 2014–2017.

Figure 11.8 shows the pictorial representation of road accidents for different states for the years 2014–2017 presented in Tables 11.7a and 11.7b. The maximum numbers of road accidents were found to be in Tamil Nadu for all the years with 77 725 (2014), 79 746 (2015), 82 163 (2016), and 74 571 (2017), whereas the minimum accidents were found in Lakshadweep.

Table 11.6 The occurrence of RTAs estimation [8].

Variable	OR adjusted	P-value	95% CI
Light condition (day)			
Night	1.35	0.005	1.1–1.7
Daylight	1.05	0.922	0.4–2.9
Sunset	2.31	0.002	1.4–3.9
Weather state (clear)			
Rainy	0.76	0.212	0.5–1.2
Cloudy	2.60	0.001	1.5–4.6
Road (highway)			
Avenue	9.74	0.001	3.0–35.5
Main road	2.84	0.120	0.8–10.6
Sidetrack	3.45	0.068	0.9–13.6
Accident site (road)			
Verge	0.90	0.635	0.6–1.4
Road edge	1.35	0.333	1.1–1.8
Area (flat)			
Mountain	1.14	0.294	0.9–1.5
Visual obstruction (no)			
Yes	0.39	0.002	0.2–0.7
Accident's place (inside the city)			
Suburb	0.95	0.786	0.6–1.3
Total	998	908	90

From the graph of Figure 11.9 for road accidents for the year 2017, we can note that the maximum value is observed in the state of Uttar Pradesh with 17 706 for fatal accidents, Kerala with 27 034 for grievous accidents, Tamil Nadu with 43 856 for minor accidents, and Maharashtra with 5216 for non-injury accidents. The minimum value is observed for Lakshadweep for all types of accidents. In Figure 11.10, the graph shows that the maximum numbers of persons killed in road accidents on national highways are in Uttar Pradesh with 5827 (2014), 7773 (2015), 7467 (2016), and 7946 (2017). The minimum is found in Dadra and Nagar Haveli and Lakshadweep.

11.5 Confidence Interval for Machine Learning—Analysis of Road Accidents

Most of the machine learning algorithm estimates are based on the performance of unseen data. Confidence interval (CI) is one such way of quantifying the uncertainty of an estimate. CI comes from the field of estimation statistics and is used to estimate population

Table 11.7a State-wise road accident data for the years 2014–2017.

	Total number of persons injured							
	State-wise				State's share			
State	**2014**	**2015**	**2016**	**2017**	**2014**	**2015**	**2016**	**2017**
Mizoram	234	103	68	55	0	0	0	0
Nagaland	230	74	120	375	0	0	0	0.1
Dadra and Nagar Haveli	96	97	130	60	0	0	0	0
Daman and Diu	49	64	102	70	0	0	0	0
Lakshadweep	1	3	0	1	0	0	0	0
Arunachal Pradesh	308	359	391	316	0.1	0.1	0.1	0.1
Meghalaya	311	319	354	354	0.1	0.1	0.1	0.1
Sikkim	352	337	263	479	0.1	0.1	0.1	0.1
Andaman and Nicobar Islands	283	331	323	263	0.1	0.1	0.1	0.1
Chandigarh	335	331	329	302	0.1	0.1	0.1	0.1
Tripura	1225	1028	853	718	0.2	0.2	0.2	0.2
Manipur	1295	1201	955	1027	0.3	0.2	0.2	0.2
Uttarakhand	1531	1657	1735	1631	0.3	0.3	0.4	0.3
Goa	1879	2055	2026	1922	0.4	0.4	0.4	0.4
Punjab	4127	4414	4351	4218	0.8	0.9	0.9	0.9
Jharkhand	4356	4038	3793	3918	0.9	0.8	0.8	0.8
Himachal Pradesh	5576	5108	5764	5452	1.1	1	1.2	1.2
Assam	6499	7068	6127	6163	1.3	1.4	1.2	1.3
Bihar	6640	6835	5651	6014	1.3	1.4	1.1	1.3
Jammu and Kashmir	8043	8142	7692	7419	1.6	1.6	1.6	1.6
Delhi	8283	8258	7154	6604	1.7	1.7	1.4	1.4
Haryana	8944	10 794	10 531	10 339	1.8	2.2	2.1	2.2
Odisha	11 087	11 825	11 312	11 198	2.2	2.4	2.3	2.4
West Bengal	12 018	11 794	11 859	10 091	2.4	2.4	2.4	2.1
Chhattisgarh	13 157	13 426	12 955	12 550	2.7	2.7	2.6	2.7
Telangana	21 636	22 948	24 217	23 990	4.4	4.6	4.9	5.1
Uttar Pradesh	22 337	23 205	25 096	27 494	4.5	4.6	5.1	5.8
Gujarat	22 493	21 448	19 949	16 802	4.6	4.3	4	3.6
Rajasthan	27 453	26 153	24 103	22 071	5.6	5.2	4.9	4.7
Andhra Pradesh	29 931	29 439	30 051	27 475	6.1	5.9	6.1	5.8

(*Continued*)

Table 11.7a (*Continued*)

	Total number of persons injured							
	State-wise				State's share			
Maharashtra	40 455	39 606	35 884	32 128	8.2	7.9	7.3	6.8
Kerala	41 096	43 735	44 108	42 671	8.3	8.7	8.9	9.1
Madhya Pradesh	55 335	55 815	57 873	57 532	11.2	11.2	11.7	12.2
Karnataka	56 831	56 971	54 556	52 961	11.5	11.4	11	11.2
Tamil Nadu	77 725	79 746	82 163	74 571	15.8	15.9	16.6	15.8

Table 11.7b State-wise road accident data for the years 2014–2017.

Total number of persons			
2014	2015	2016	2017
34.4	33.6	34	30.9
24	27.6	29.8	23.8
20.5	22	18.9	18.8
6.5	6.7	5.4	5.7
52.1	52.5	50.1	47.9
98.1	105.2	102.5	95
36.7	34.5	31.8	26.4
33.5	39.9	38.3	37.1
79.9	72.6	81.2	76.2
66.2	66.3	61.9	59.1
13.3	12.2	11.3	11.5
92.8	92.2	87.5	84.2
116.6	123.3	123.6	119
73.2	72.7	74.3	72.9
34.5	33.4	29.9	26.4
51.1	46.9	36.8	39.2
11.5	11.6	12.8	12.6
22.5	9.8	6.4	5.1
9.9	3.1	5	15.6
26.5	28.1	26.6	26.2
14.4	15.3	14.9	14.4
38.7	36.3	33	29.9
55.6	52.7	40.6	73.4
113.2	115.5	118.4	106.9
NA	NA	NA	NA
32.7	27.2	22.3	18.6

(*Continued*)

Table 11.7b (*Continued*)

Total number of persons			
14.8	15.8	16.3	15.2
10.6	10.8	11.5	12.4
13.1	12.7	12.7	10.7
53.1	61	58.6	46.7
20.3	19.3	18.5	16.2
23.9	23.4	30.8	13.7
16.1	20.2	30.9	20.3
41.2	39.9	33.6	30.2
1.3	3.8	0	1.2

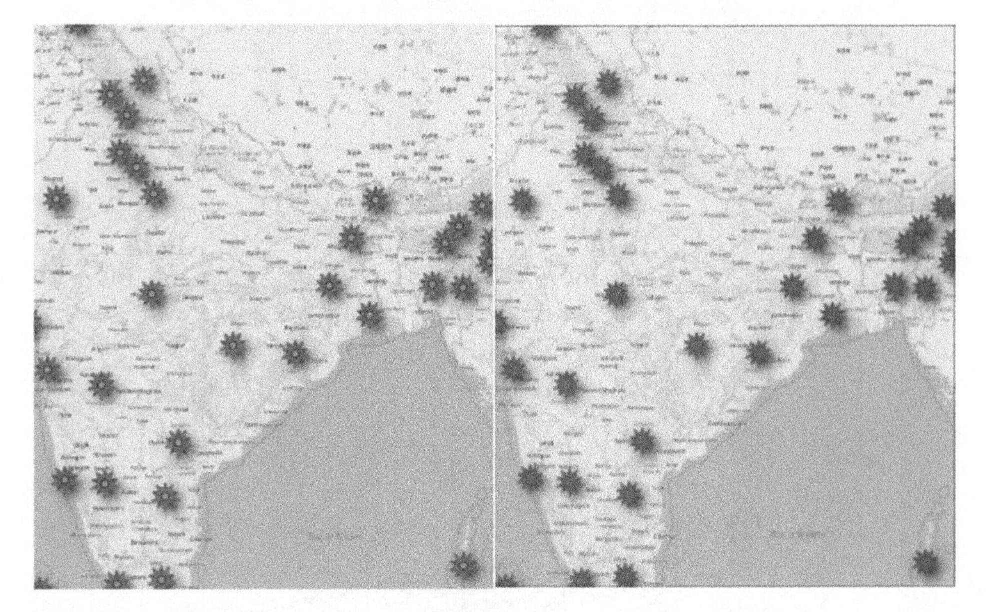

Figure 11.7 Road accidents for the different states for the year 2017: (a) fatal and grievous injury accidents, (b) minor injury and non-injury accidents.

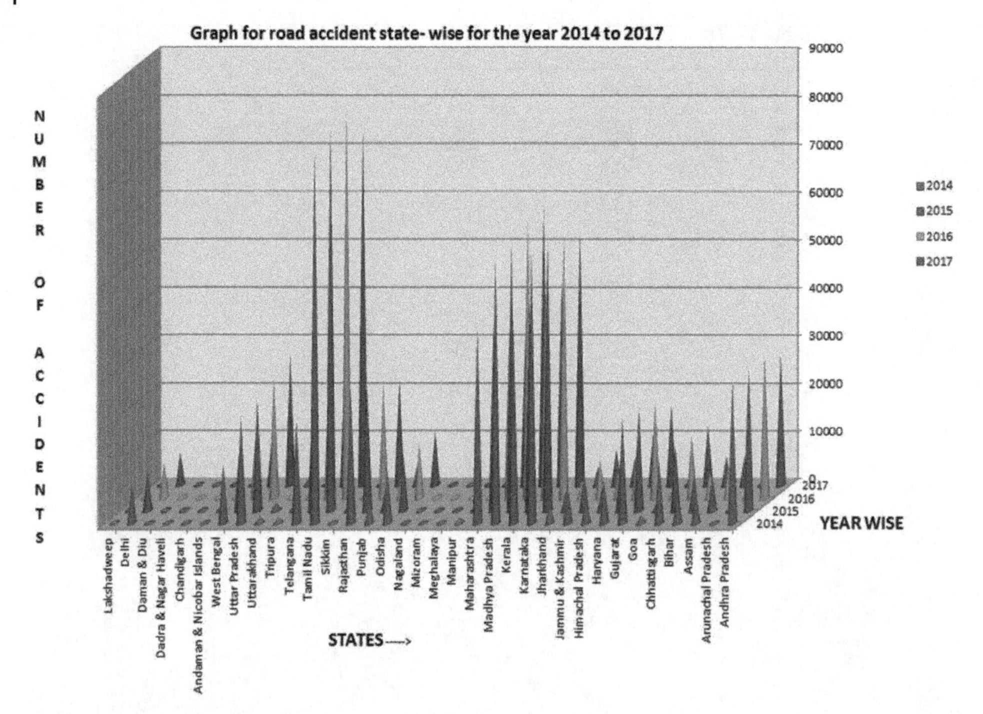

Figure 11.8 Graph for road accidents for different states for the years 2014–2017.

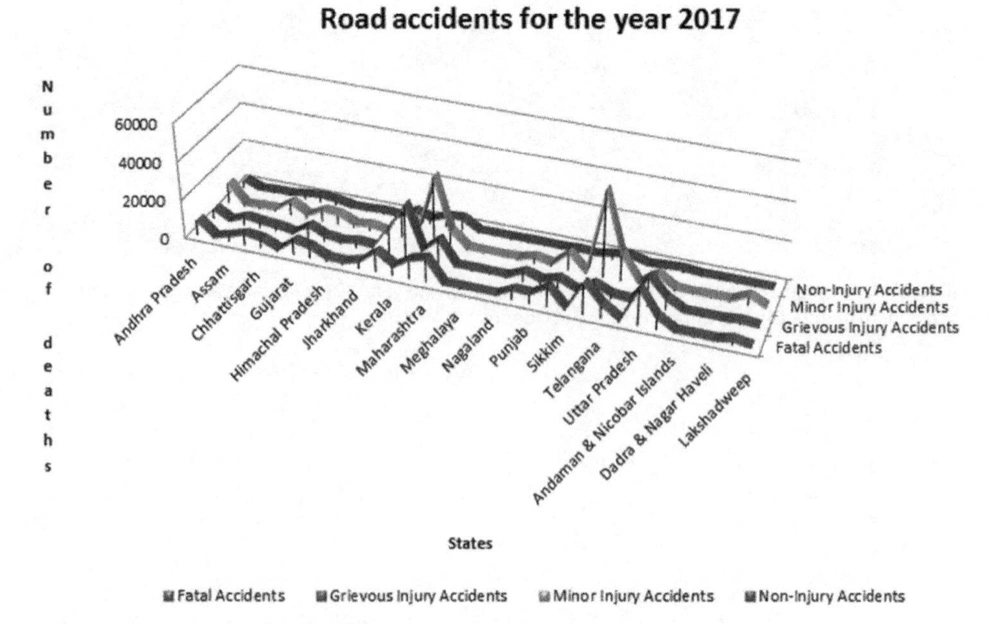

Figure 11.9 Graph for road accidents for different states for the year 2017.

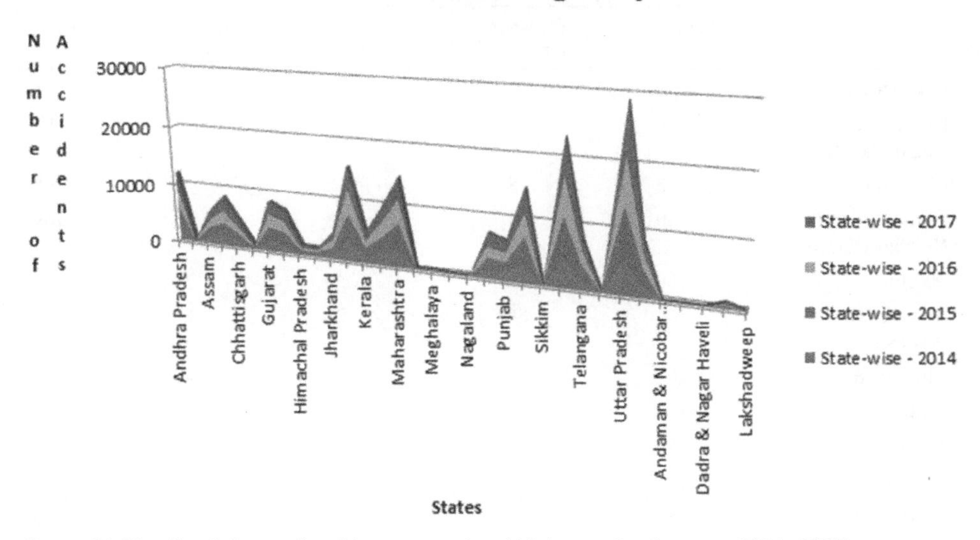

Figure 11.10 Graph for road accidents on national highways for the years 2014–2017.

parameters and arbitrary population statistics. Estimation statistics describes three classes, namely: (i) the *effect size*, which quantifies the size of an effect given intervention, (ii) *interval estimation*, which quantifies the amount of uncertainty in a value, and (iii) *meta-analysis*, which quantifies the findings across multiple similar studies.

Suppose 95% confidence is between 80% (84 − 4) and 88% (84 + 4). Then, the CI is given as the interval 84 4, and the confidence is 95%, whereas the actual percentage is 84%.

11.5.1 Results of Confidence Interval

The CI is calculated using STATA software for Tables 11.7 and 11.8. The 95% CI is between 7447.53% and 20 887.8% for the column "State-wise total number of persons injured in road accidents," and 95% CI is between 1.58% and 4.43% for the column "Share of states—total number of persons injured in road accidents." The calculated mean and standard error are 14 167.67 and 3299.14 for columns (SWRA), respectively.

The 95% CI for RA in the year 2016 is found to be 7758.83%–21 998.81%, with mean 14 878.82 and standard error 3495.44. The 95% CI (2015) is found to be 8005.67%–22 115.9% with mean 15 060.79 and standard error 3463.59, and 95% CI (2014) is found to be 7228.21%–20 187.01% with mean 13 707.61 and standard error 3191.65, as shown in Tables 11.9–11.12.

CI for Table 11.8 is shown in Table 11.13 for the year 2017; 95% CI for fatal accidents is 2406.96%–5702.85%, 95% CI for grievous accidents is 1599.03%–5654.84%, 95% CI for minor accidents is 1833.28%–8677.14%, and 95% CI for non-injury accidents is 525.07%–1572.62%.

Table 11.8 Road accidents for different states for the year 2017.

	Accidents				
States	**Fatal**	**Grievous injury**	**Minor injury**	**Non-injury**	**Total**
Lakshadweep	0	1	0	0	1
Dadra and Nagar Haveli	40	20	4	3	67
Mizoram	55	8	2	3	68
Andaman and Nicobar Islands	20	54	88	27	189
Sikkim	60	70	58	8	196
Arunachal Pradesh	103	86	26	26	241
Chandigarh	103	7	198	34	342
Tripura	153	339	2	9	503
Nagaland	35	72	159	265	531
Manipur	107	91	342	38	578
Meghalaya	140	253	101	181	675
Himachal Pradesh	907	959	1042	206	3114
Goa	306	237	926	2448	3917
Jharkhand	3034	1734	337	93	5198
Jammu and Kashmir	765	2179	1945	735	5624
Punjab	4139	1490	561	83	6273
Delhi	1565	907	4110	91	6673
Assam	2474	3451	706	539	7170
Bihar	5045	2431	887	492	8855
Odisha	4372	4021	2302	160	10 855
Haryana	4700	1700	4771	87	11 258
West Bengal	5199	4811	560	1061	11 631
Chhattisgarh	3878	1706	6285	1694	13 563
Gujarat	6739	5653	5033	1656	19 081
Rajasthan	9300	4017	8110	685	22 112
Telangana	6110	1165	12 695	2514	22 484
Andhra Pradesh	7564	4607	10 285	3271	25 727
Maharashtra	11 220	12 164	7253	5216	35 853
Kerala	3915	27 034	5994	1527	38 470
Uttar Pradesh	17 706	14 363	6044	670	38 783
Karnataka	9739	14 191	14 247	4365	42 542
Madhya Pradesh	9258	4863	34 493	4785	53 399
Tamil Nadu	15 061	5005	43 856	1640	65 562

Table 11.9 CI for road accidents in the year 2017.

Variable	Obs.	M	SE	95% CI
SWRA (2017)	33	14 167.67	3299.14	7447.53–20 887.8
TPI	33	3.01	0.69	1.58–4.43

SWRA: State-wise total number of persons injured in road accidents TPI: Share of states–total number of persons injured in road accidents. M: mean; SE: standard error.

Table 11.10 CI for road accidents in the year 2016.

Variable	Obs.	M	SE	95% CI
SWRA (2016)	33	14 878.82	3495.44	7758.83–21 998.81

Table 11.11 CI for road accidents in the year 2015.

Variable	Obs.	M	SE	95% CI
SWRA (2015)	33	15 060.79	3463.59	8005.67–22 115.9
TPI	33	3.01	0.69	1.60–4.42

Table 11.12 CI for road accidents in the year 2014.

Variable	Obs.	M	SE	95% CI
SWRA (2014)	33	13 707.61	3191.65	7228.21–20 187.01
TPI	33	2.84	0.66	1.50–4.194

Table 11.13 CI for the year 2017.

Variable	Obs.	M	SE	95% CI
FA	33	4054.90	809.03	2406.96–5702.85
GA	33	3626.93	995.56	1599.03–5654.84
MIA	33	5255.21	1679.94	1833.28–8677.14
NIA	33	1048.84	257.13	525.07–1572.62

FA: fatal accidents; GA: grievous accidents; MIA: minor injury accidents; NIA: non-injury accidents.

From all the tables, the highest 95% CI (8005.67–22 115.9) was found to be in the year 2015, and a drastic rise in the number of accidents was found in the states of Madhya Pradesh, Telangana, and Uttar Pradesh. A decline was found in many states including Andhra Pradesh, Assam, Gujarat, Karnataka, and Tamil Nadu, to name a few.

11.6 Deep Learning in Traffic Management

One of the most significant roles of traffic management is identifying accidents and alerting the police and public, and also controlling the traffic for smooth functioning on roads, highways, and so on.

Acharya et al. [9] implemented a framework detection method for a robust parking lot using deep convolutional neural network (CNN) and a binary support vector machine classifier. The classifier was trained and tested by the features learned by deep CNN from public datasets (PK lot) having different illuminance and at different weather conditions. This in turn would help prevent traffic congestion by identifying beforehand where to park the vehicle and conducting the flow of traffic smoothly. The results achieved a high accuracy of 99.7% on the training dataset and a transfer learning accuracy of 96.6% on an independent test dataset (see Figure 11.11), which indicates its suitability for mass applications in all weather conditions. The framework provides a cheap and reliable solution.

Figure 11.11 (a) Car parking lot, (b) car is placed, and (c) empty space available for parking.

Deep learning techniques are used everywhere and can be implemented in traffic management too; this will help smart roads and, thereafter, smart cities too. In this study, a few images of areas having traffic jams were obtained, and the bounding boxes technique was implemented on these images. Object detection using the YOLO algorithm for bounding boxes in Python produced the results shown in Figures 11.12 and 11.13. This helps to demonstrate that traffic management can use deep learning techniques to identify any accident or any kind of disturbance that occurs on the road, and helps to analyze why and what causes traffic congestion.

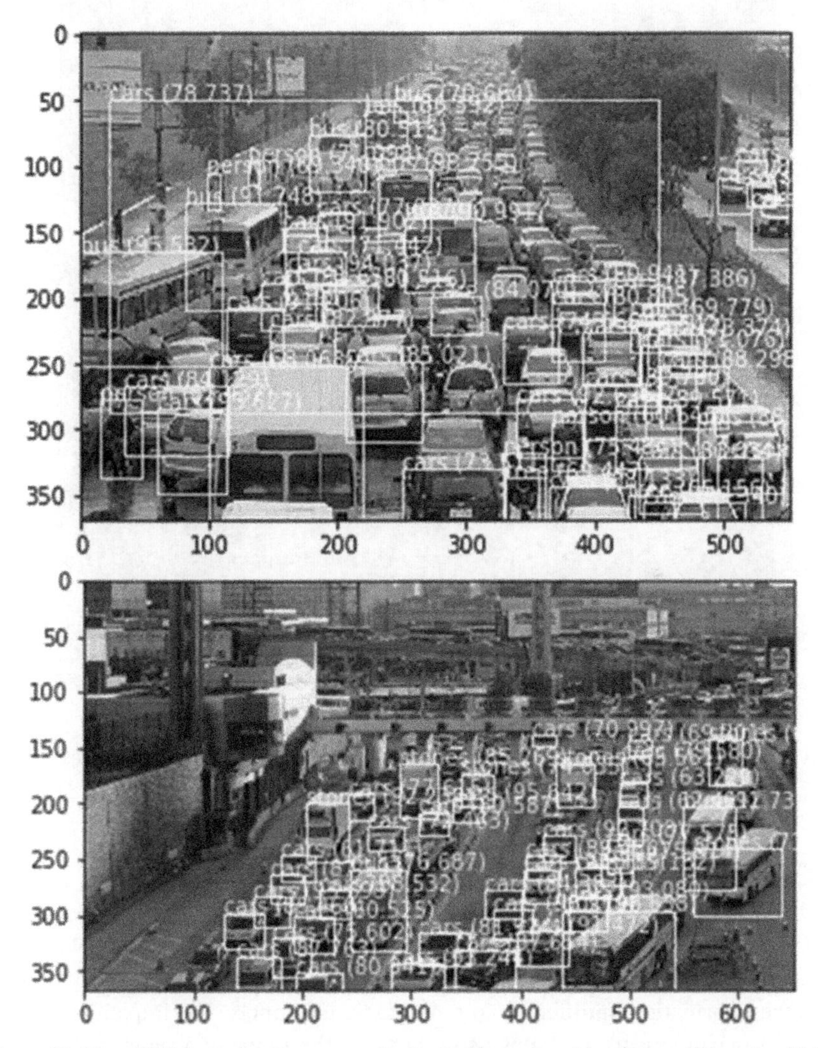

Figure 11.12 YOLO bounding boxes results for traffic system identifying different vehicles.

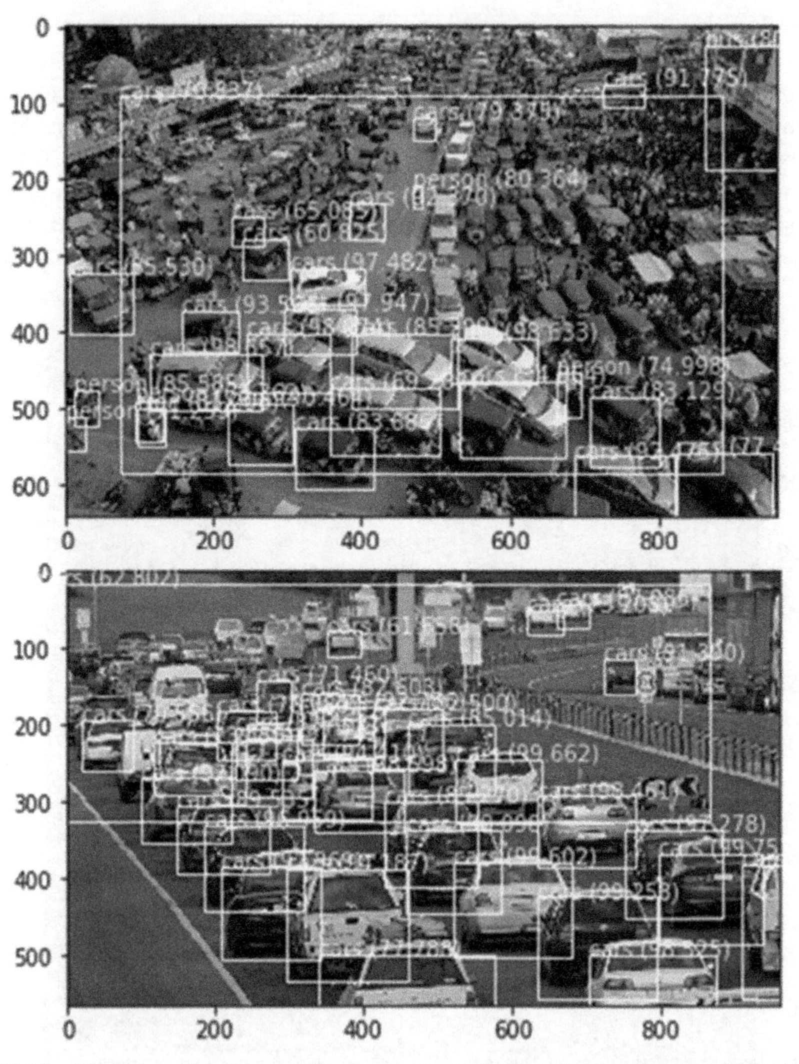

Figure 11.13 YOLO algorithm for bounding boxes results for identifying different vehicles in traffic.

11.7 Conclusion

This study concludes that traffic management is essential for the smooth flow of traffic in cities. The traffic control algorithm works on the principle of scheduling algorithms and exchanges the information in different communication channels. The implementation of GIS and CI of machine learning helps us understand the range of accidents that occur. Hence, TMS can be implemented by deep learning techniques. This makes the city easy to monitor and integrate with the conditions of all infrastructures, including roads, bridges, tunnels, railways, metros, airports, and seaports. The communication, water, and power systems can be better analyzed using this approach. It can be used to reduce the need for maintenance activities and increase monitoring security for citizen services. The key goal of smart roads in smart cities is to improve the quality of living using informatics and technology. The efficient services rendered will satisfy the needs of the citizens. This reduces the overall pollution and general environmental impact and smoothens traffic flow.

References

[1] R.A. Eskesen, J.T. Grundahl, and C.D. Jensen, "Smart City Traffic Management," Technical University of Denmark, 2017.

[2] M. Pop and O. Proştean, "A comparison between smart city approaches in road traffic management," *Procedia – Soc. Behav. Sci.*, vol. 238, pp. 29–36, 2018.

[3] A.N.N. Kishore and Z. Sodhi, *Exploratory Research on Smart Cities. Theory, Policy and Practice*, Pearl, New Delhi, 2015.

[4] S. Devi and T. Neetha, "Machine Learning based traffic congestion prediction in a IoT based Smart City," *Int. Res. J. Eng. Techn. (IRJET)*, vol. 04, no. 5, pp. 3442–3445, 2017.

[5] A. Khanna, R. Goyal, M. Verma, and D. Joshi, "Intelligent traffic management system for smart cities," In *Futuristic Trends in Network and Communication Technologies* (pp. 152–164), 2018.

[6] J.J. Rolison, S. Regev, S. Moutari, and A. Feeney, "What are the factors that contribute to road accidents? An assessment of law enforcement views, ordinary drivers' opinions, and road accident records," *Accid. Anal. Prev.*, vol. 115, pp. 11–24, 2018.

[7] A. Sami, G. Moafian, A. Najafi, M. RezaAghabeigi, N. Yamini, S.T. Heydari, and K.B. Lankarani, "Educational level and age as contributing factors to road traffic accidents," *Chin. J. Traumatol.*, vol. 16, no. 5, pp. 281–285, 2013.

[8] M.M. Jalilian, H. Safarpour, J. Bazyar, S.M. Keykaleh, L. Malekyan, and A. Khorshidi, "Environmental related risk factors to road traffic accidents in Ilam, Iran," *Med Arch*, vol. 73, no. 3, pp. 169–172, 2019.

[9] D. Acharya, W. Yan, and K. Khoshelham (2018). Real-time image-based parking occupancy detection using deep learning. In *Proceedings of the 5th Annual Conference of Research@ Locate, Adelaide, Australia*, CEUR Workshop Proceedings, vol. 2087, pp. 33–40.

Index

Sensor Data Analysis and Management: The Role of Deep Learning, First Edition. Edited by A. Suresh, R. Udendhran, and M.S. Irfan Ahmed.
© 2021 John Wiley & Sons, Ltd. Published 2021 by John Wiley & Sons, Ltd.